Wael Alkasawneh

Backcalculation of Pavement Moduli Using Genetic Algorithms

Wael Alkasawneh

Backcalculation of Pavement Moduli Using Genetic Algorithms

State of the Art Methods and Techniques

VDM Verlag Dr. Müller

Imprint

Bibliographic information by the German National Library: The German National Library lists this publication at the German National Bibliography; detailed bibliographic information is available on the Internet at http://dnb.d-nb.de.

Cover image: www.purestockx.com

Publisher:
VDM Verlag Dr. Müller Aktiengesellschaft & Co. KG, Dudweiler Landstr. 125 a, 66123 Saarbrücken, Germany,
Phone +49 681 9100-698, Fax +49 681 9100-988,
Email: info@vdm-verlag.de

Produced in USA and UK by:
Lightning Source Inc., La Vergne, Tennessee, USA
Lightning Source UK Ltd., Milton Keynes, UK
BookSurge LLC, 5341 Dorchester Road, Suite 16, North Charleston, SC 29418, USA

ISBN: 978-3-639-00477-9

TABLE OF CONTENTS

ii

LIST OF TABLES

vii

LIST OF FIGURES

xi

CHAPTER I

INTRODUCTION

1.1 Statement of the Problem

The backcalculation of pavement moduli from surface deflection measurements using nondestructive tests has been used for more than four decades to assess and manage existing pavements and to design overlays. Unfortunately, the backcalculated pavement moduli lack the accuracy in spite of the existence of many backcalculation programs employing different backcalculation procedures and algorithms.

The value and accuracy of the backcalculated moduli are highly dependent on the backcalculation procedure (Chou and Lytton, 1991; Harichandran et *al.*, 1993). All of the classical backcalculation procedures require seed moduli to initiate the backcalculation process. Different seed moduli often lead to different backcalculated moduli which in turn lead to different pavement designs and evaluations adding more challenges to engineers.

The main problems that any classical backcalculation procedure faces are convergence, accuracy, and the number of layers in the backcalculation program. The selection of the seed moduli controls the convergence of the backcalculation procedure to pavement moduli that minimizes the mean square error (objective function) between the measured deflection basin and the backcalculated deflection basin using the backcalculated moduli. It is known that more than one solution can satisfy the objective function criterion in backcalculating the pavement moduli due to the multimodal nature of the backcalculation search space where many local optima exist. In turn, the arrival at local optima will lead to "inaccurate" pavement moduli that can be as twice as the "accurate" value. On the other hand, the maximum number of layers than can be used in any backcalculation program does not exceed 5 layers with recommendations to use 3 layers to reduce the error associated with the backcalculation process. In some cases, increasing the number of the layers in the backcalculation process is desirable to obtain more representative variation of the moduli with depth.

1

Genetic algorithms (GAs) can be used to backcalculate the pavement moduli by searching the entire search space of the objective function using guided random search techniques. The GAs is based on the Darwinian theory and formulated on the mechanics of genetics and natural selection (Holland, 1975; Goldberg, 1989). In nature, species "adapt" themselves, under the effect of the surrounding environment and conditions, for survival. In GAs, randomly generated solutions are used to generate a pool of feasible solutions that "evolve" from one generation to another using algorithms that mimic the natural selection in nature such as mutation and crossover (reproduction).

Very few researches have been performed to study the backcalculation of the pavement moduli using GAs (Fwa et al., 1997; Kameyama et al., 1997; Reddy et al., 2004; Tsai et al., 2004). These researches were geared toward studying the feasibility of using the GAs method as a backcalculation procedure and toward finding "appropriate" values for the GAs parameters. The researches were successful at pointing out that the GAs method is a robust and accurate method to backcalculate the pavement moduli using measured deflections. However, the researches failed to select the appropriate GAs parameters due to the complexity of the search space which is dependent on the forward calculation procedure, the number of pavement layers, the ranges of the backcalculated moduli, and the used objective function. In addition, the interaction between the genetic parameters themselves and the effect of the genetic parameters and operators on the performance of the backcalculation process add more complexity to the backcalculation process. Therefore, the GAs method is not used widely since it is not designed, yet, to be used by practitioners rather than researchers and more research is still ongoing to study the many factors that affect the backcalculation process.

The lack of enough research and guidelines on how to use the GAs in the backcalculation of the pavement moduli stimulates the need to develop a robust backcalculation program that is user friendly and at the same time accurate. This can be done by identifying the main factors that affect the GAs performance and by studying thoroughly the performance of the backcalculation procedure. In addition, a successful procedure that can be used in practice should be easy to use and powerful enough to eliminate all the drawbacks associated with other classical backcalculation procedures and programs. The new backcalculation program (BackGenetic3D) is

2

based on the MultiSmart3D forward calculation program developed by the Computer Modeling and Simulation Group at the University of Akron (Pan and Alkasawneh, 2006).

1.2 Objectives

The objectives of this research are to:

1) Study the uniqueness and effect of the root mean square error as an objective function to backcalculate the static pavement moduli.

2) Study the complexity of the search space of the backcalculated moduli.

3) Study the factors that affect the genetic algorithms performance when used as a backcalculation procedure.

4) Study the interaction between the genetic algorithms parameters and operators.

5) Select a new objective function to enhance the backcalculation performance.

6) Develop a new backcalculation program (BackGeneitc3D) that is user friendly implementing more accurate genetic operators.

7) Generalize the backcalculation procedure to include any arbitrary number of pavement layers, loading conditions, loading configuration, and number of sensors.

8) Automate the backcalculation process by eliminating the dependency of the program on the user's inputs (Parameterless Genetic Algorithms-PGA).

8) Develop a new PGA that can be used in the backcalculation of the pavement moduli and any other genetic algorithms application.

9) Validate the performance and accuracy of the new backcalculation program using several pavement cases.

1.3 Scope of the Book

The work involved in this book can be grouped into two parts. The first part includes studying the performance of the genetic algorithms thoroughly by identifying all the factors that can affect the backcalculation procedure. The second part includes the development of a robust backcalculation program that is both accurate and user friendly. In addition, the second part includes the automation of the backcalculation procedure by reducing the number of user's inputs and hence eliminating the need for a user who is well experienced with genetic algorithms.

3

A literature review on the backcalculation methods of pavement layer moduli is performed. The limitations and drawbacks of each backcalculation method are identified and discussed. All of the genetic parameters and operators were studied thoroughly using different examples. In addition, 3-, 4-, and 23-layer pavement systems were analyzed and backcalculated to study the mechanism of the backcalculation procedure and the performance of the genetic algorithms.

A new program was developed using the GAs concepts where new genetic operators and parameters were introduced for the first time to backcalculate the pavement moduli. The new program (BackGenetic3D) can be used to backcalculate the static pavement moduli for any number of layer, loading conditions, and loading configuration. Current classical backcalculation programs are limited to one circular load, 5-layer system, and uniform loading only. The BackGenetic3D program assumes that the pavement layers can be analyzed assuming homogeneous, linear, and elastic materials.

A new powerful Dynamic Parameterless Genetic Algorithm (DPGA) was developed based on the PGA developed by Harik and Lobo (1999). The new DPGA eliminates the need for the GAs parameters by automating them. The new method is very powerful and robust and has been studied using a 23-layer pavement system.

1.4 Outlines of the Dissertation

Chapter II presents the literature review on forward and backcalculation analysis methods of pavement systems assuming homogeneous, linear, and elastic materials. In addition, the limitations of the backcalculation methods are discussed.

Chapter III studies the uniqueness of the backcalculated pavement moduli using the root mean square error as the objective function. The effects of the root mean square error and the backcalculated moduli on the rutting and fatigue in flexible pavements are investigated as well.

Chapter IV reviews the fundamentals of genetic algorithms and the main used parameters and operators. In addition, the interaction between the operators and parameters are discussed.

Chapter V presents the main features and the techniques implemented in the BackGenetic3D program. The program was verified using results from other classical backcalculation programs.

Chapter VI introduces a new method to select the genetic parameters. Large and small population sizes were used to backcalculate the pavement moduli of a 3-layer pavement system. The performance of the backcalculation process was discussed thoroughly. In addition, guidelines were provided on the selection of the moduli ranges for the backcalculation process.

Chapter VII investigates the effect the chromosome length, the fitness function, the number of layers, and the number of surface deflection points (sensors) on the performance of the backcalculation process. Examples using 3- and 23-layer pavement systems were analyzed. Guidelines and recommendations were provided as well.

Chapter VIII introduces new innovative methods to automate the backcalculation process. The Parameterless Genetic Algorithm (PGA) and the Dynamic Parameterless Genetic Algorithm (DPGA) were introduced and compared using a 23-layer pavement system. Guidelines and recommendations on the use of the PGA and the DPGA are provided as well.

Chapter IX presents summaries and conclusions of the research work.

CHAPTER II

PAVEMENT ANALYSIS

2.1 Introduction

Roads are very important in everyday transportation activities. Durability of pavement and driving convenience are very essential in pavement design and rehabilitation. The cost of pavement material and design is largely affected by the material type which can range widely based on the roadway design parameters. A maintenance-free pavement is still difficult and more research is needed to find an economical maintenance-free pavement.

Performance and durability of existing and new pavements are very important to increase the design life of pavements and to reduce the roadway hazards. In the US, a big portion of the transportation agencies budget is spent every year on pavement maintenance and improvement. In Ohio, for example, the length of public roads increased from approximately 117,000 miles to 123,000 miles between 2000 and 2003. However, the design procedures and practice depended largely on procedures and practices that are more than 50 years old. It can be argued that the need for more practical and up-to-date procedures is inevitable these days than ever due to the complexity of roadways design and the advancement in both technology and behavioral mechanisms of pavements.

New design approaches and procedures have been developed recently by The National Cooperative Highway Research Program (NCHRP) and the American Association of State Highway and Transportation Officials (AASHTO) Joint Task Force on Pavements. The new effort has resulted in replacing the old empirical-based pavement design procedures with mechanistic-empirical (M-E) based procedures in the new Mechanistic-Empirical Pavement Design Guide (MEPDG). The M-E approach is a very powerful approach since it combines the actual observed behavior of pavements and the analytical modeling techniques that are widely accepted. The M-E procedures are more reasonable and realistic than the empirical-based procedures since they allow for more consideration of the effects of the actual site traffic distribution, climate, material types, structure, and other design features.

Pavement behavior is a key point in the new MEPDG guide. Pavement behavior and empirical procedures have been developed based on laboratory and full scale field tests that provide a valuable data base for analytical procedures. Transportation agencies under the new MEPDG are given more freedom to tailor the design procedures to suit their roads and pavement conditions.

The M-E procedures can be considered as new design procedures and can be used to verify the design parameters with the actual parameters including traffic loading, material properties, and climatic conditions.

Understanding the pavement structural responses along the pavement section and the load capacity of the pavement requires suitable theories that represent the actual mechanical behavior. The following discusses some of the available theories that have been used to simulate pavement structural responses.

2.2 Elasticity Theory

Elasticity theory was the first theory to be used for pavement analysis. Applying this theory to study the structural response of the pavement assumes that the pavement material is elastic. Elasticity theory works well as long as the stress-strain ratio is constant. This indicates that the elasticity theory is well suited for pavement sections that do not undergo stresses greater than the failure stresses. Equations derived from elasticity theory use the same basic theory but different assumptions for material properties and geometry (Bendana et. al, 1994). Pavement analysis using the elasticity theory is normally performed using Hooke's law and Boussinesq theory.

2.2.1 Hooke's Law

Hooke's law is based on the assumption that the stress-strain ratio is constant in the uniaxial load case of the material. Hooke's law was derived assuming that the material is perfectly elastic and homogeneous.

2.2.2 Boussinesq's Equation

Boussinesq developed equations to compute stresses within a homogeneous, isotropic, linearly elastic half space under a point load acting perpendicular to the surface. The half-space assumption indicates an infinitely large area and an infinite depth. The value of the stress is given by (Holtz and Kovacs, 1981):

$$\sigma_z = \frac{P(3z^3)}{2\pi(r^2+z^2)^{5/2}} = \frac{P(3/2\pi)}{z^2(1+\frac{r^2}{z^2})^{5/2}}$$
                                                              Eq. (2.1)

where P=point load, z=depth from ground surface to the stress point, r=horizontal distance from the point load to the stress point.

Boussinesq developed other equations to compute the state of shear stresses, normal strains, and displacements under a point load in the elastic material as shown in Table 2.1 and Figure 2.1. It can be seen from the table that the normal strains, displacements, tangentional stresses, and radial stresses depend on the Poisson's ratio ν and/or the modulus of elasticity E while the vertical stress and shear stresses are independent of the Poisson's ratio and the modulus of elasticity.

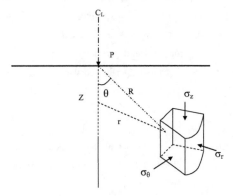

Figure 2.1, Notation for Boussinesq's equation (Ullidtz, 1987)

8

Table 2.1, Boussinesq's Equations for a Point Load

Normal Stresses

$$\sigma_z = \frac{3P}{(2\pi R^2)}\cos^3\theta$$

$$\sigma_r = \frac{P}{(2\pi R^2)}\left(3\cos\theta\sin^2\theta - \left(\frac{1-2v}{1+\cos\theta}\right)\right)$$

$$\sigma_t = \frac{P}{(2\pi R^2)}(1-2v)\left(-\cos\theta + \left(\frac{1}{1+\cos\theta}\right)\right)$$

$$\sigma_v = \frac{1}{3}(\sigma_z + \sigma_r + \sigma_t) = \frac{P}{(3\pi R^2)}(1+v)\cos\theta$$

Shear Stresses

$$\tau_{rz} = \frac{3P}{(2\pi R^2)}\cos^2\theta\sin\theta$$

$$\tau_{rt} = \tau_{tz} = 0$$

Normal Strains

$$\varepsilon_z = \frac{(1+v)P}{(2\pi ER^2)}\left(3\cos^3\theta - 2v\cos\theta\right)$$

$$\varepsilon_r = \frac{(1+v)P}{(2\pi ER^2)}\left(-3\cos^3\theta + (3-2v)\cos\theta - \left(\frac{1-2v}{1+\cos\theta}\right)\right)$$

$$\varepsilon_t = \frac{(1+v)P}{(2\pi ER^2)}\left(-\cos\theta + \left(\frac{1-2v}{1+\cos\theta}\right)\right)$$

$$\varepsilon_v = \varepsilon_z + \varepsilon_r + \varepsilon_t = \frac{(1+v)P}{(\pi ER^2)}(1-2v)\cos\theta$$

Displacements

$$d_z = \frac{(1+v)P}{(2\pi ER)}\left(2(1-v) + \cos^2\theta\right)$$

$$d_r = \frac{(1+v)P}{(2\pi ER)}\left(\cos\theta\sin\theta - \frac{(1-2v)\sin\theta}{1+\cos\theta}\right)$$

$$d_t = 0$$

Boussinesq's equation can be extended to other loading conditions such as line load over a finite area (Newmark's equation; Newmark, 1935). Westergaard (1938) derived an equation for stresses under a point load in elastic homogeneous half-spaces with a Poisson's ratio equal to zero. His equation is given in Eq. (2.2) where the terms are similar to the terms in Boussinesq's equation. Both Boussinesq's and Westergaard's equations provide almost the same stress value for $r/z \geq 1.5$, while for $r/z < 1.5$ Boussinesq's equation provides larger values than Westergaard's equation (Holtz and Kovacs, 1981).

Boussinesq (1876, 1885) developed a more realistic theory for stresses in granular material assuming a variable shear modulus that changes as the confining stresses change. This assumption resembles actual stresses in materials where the shear modulus increases as the stresses increase.

Elasticity theory and Boussinesq's equations were studied by other researchers to check the validity of the theory and the accuracy of the assumptions. Frolich (1934) showed that the radial stress in Boussinesq's theory is the major principal stress which is inversely proportional to the square of the distance from the point load. Frolich (1934) generalized the theory by introducing a concentration factor (n) as shown in the following equations.

$$\sigma_z = \frac{nP}{(2\pi R^2)} \cos^n \theta \qquad \text{Eq. (2.2)}$$

$$\sigma_r = \frac{nP}{(2\pi R^2)} \cos^{n-2} \theta \sin^2 \theta \qquad \text{Eq. (2.3)}$$

$$\sigma_t = 0 \qquad \text{Eq. (2.4)}$$

$$\sigma_R = \frac{nP}{(2\pi R^2)} \cos^{n-2} \theta \qquad \text{Eq. (2.5)}$$

$$P = \frac{\sigma_R}{3} \qquad \text{Eq. (2.6)}$$

If the modulus changes with the hydrostatic stress or the major principal stress is raised to a power factor, two cases arise: The first case is resulted from the assumption that the modulus changes with the hydrostatic stress which results in a positive power factor and hence a dilation of the material. The second case is resulted from the assumption that the modulus changes with the major principal stress which produces a negative power factor and hence a "stress dispersion" rather than "stress concentration" as proposed by Boussinesq's theory (Ullidtz, 1998).

2.2.3 Charts Method

Boussinesq solution was extended by different researchers to account for more general loading conditions. Solutions for a circular loaded area can be obtained by integrating the Boussinesq solution for a point load. In the influence lines method, solutions for stresses and deflections can be obtained by referring to charts that had been developed by simplifying the existing conditions under a circular loaded area and by using the elastic theory. These assumptions include neglecting the Poisson's ratio (Foster and Ahlin, 1954), or including the Poisson's ratio (Ahlvin and Ulery, 1962). The Poisson's ratio has relatively little effect on stresses and strains in the half-space elastic theory and therefore can be neglected for simplicity (Huang, 1993).

2.3 Method of Equivalent Thickness (MET)

This method was first developed by Odemark (Ullidtz, 1987). The MET method transforms a system with different moduli to a system with one modulus in order to apply the classical elasticity theories to the system. This method involves two steps in which the interface plays a key rule for the transformation. In Figure 2.2, stresses and strains above the first layer can be found by using the same modulus of elasticity and Poisson's ratio for the second layer as in the first layer and assuming a half-space case. Stresses and strains in the second layer or at the interface can be found by transforming the first layer into a layer with the same modulus of elasticity and Poisson's ratio as in the second layer but with a new thickness based on the original stiffness of the first layer.

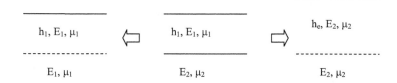

Figure 2.2, Typical transformations in the MET method, after Ullidtz (1987).

11

Theoretically, the MET method can be applied to any system with any number of layers as shown in Figure 2.3.

h_1, E_1, ν_1

h_2, E_2, ν_2

h_3, E_3, ν_3

$\vdots$

h_n, E_n, ν_n

$\vdots$

Figure 2.3, Typical multilayer system, after Ullidtz (1987).

The equivalent thickness of the transformed layer based on the original stiffness of the layer can be found using the following equation (Ullidtz, 1987):

$$h_{e,n} = f \sum_{i=1}^{n-1} \left(h_i \left[\frac{E_i}{E_n} \right]^{\frac{1}{3}} \right)$$

Eq. (2.7)

where $h_{e,n}$ is the equivalent thickness for n layers, h_i, E_i and E_n are shown in Figures 2.2 and 2.3, f is the correction factor discussed below.

The MET method is primarily based on the assumption that stress distribution below the transformed layers will be the same since an equivalent stiffness is used for each layer. Stresses and strains are assumed to be linear in each layer. However, the MET method is an approximate method and stresses and strains estimated based on this method should be corrected to improve the agreement with layered elastic theory. The correction factor should be estimated based on the number of layers in the system, Poisson's ratio, modulus of elasticity, and layer thickness. In general, the correction factor for the first interface in a two-layer system is recommended to be 0.9 while it is 1.0 in a multilayer system. The correction factor for other interfaces in a

12

multilayered system is recommended to be 0.8 (Ullidtz, 1987). The MET method can overestimate or underestimate stresses and strains in the layer and at the interface and hence can produce misleading results unless a good correction factor is available.

2.4 Finite Element Method

Finite Element Method (FEM) has been used to study the responses of flexible pavements using any material constitutive law. In the FEM method, the geometry under study is discretized into small elements connected by nodes to resemble the actual geometry or domain. Constitutive laws then applied to govern the behavior of the material, and stresses and strains can be estimated accordingly. The advantage of the FEM analysis comes from the ability to simulate any loading condition, static and dynamic, and any geometrical variation, and local discontinuities such as cracks or joints. The FEM deals with materials as continuum and therefore it simplifies the actual behavior of granular material.

Most practical FEM modeling involves only two-dimensional analysis of pavement sections due to cost, time, and modeling limitations associated with the three dimensional analysis. Three-dimensional analysis involves the discretization of the domain using sophisticated meshing techniques that adversely affect the time and cost needed to study the pavement response. Two-dimensional analysis assumes axis-symmetric equilibrium conditions that limit the simulation of the full geometry when having local discontinuities.

2.5 Discrete Element Method

Advanced numerical simulations assume continuum medium when dealing with pavement modeling, which indicates a compatibility assumption unless special considerations are allowed to account for discontinuities such as using elements in the finite element method.

Granular materials like asphalt can be described using the granular material physical behavior. Physical behavior of granular material includes normal and shear forces, and translational and rotational displacements between grains. Strains in granular materials are negligible under normal stresses.

In 1978, Cundall (1978) proposed the Distinct Element Method which uses gravity, external forces, mass of particles, center of gravity of particles, and moment of inertia of

13

particles with Newton's laws to describe the movement and interaction between grains. Discrete element simulation is carried out using steps in which forces and displacements are calculated for the medium and used as inputs for the next step.

Ullitdz (1995,1998), using circular disks to simulate grains, showed that the Distinct Element Method produces much larger tensile strains close to the axes of the load than continuum mechanics. This shows that continuum mechanics has some limitations when predicting stresses and strains in particulate media. In addition, it was shown that plastic strains are calculated simultaneously with the resilient or elastic strains, as well as strains at failure. However, this advanced simulation technique is far from being available to engineers and is still a research tool.

2.6 Visco-Elasticity Method

Materials deformation can be elastic, plastic, viscous, and viscoelastic. Based on the physical behavior, these deformations can be categorized as energy-storage processes and dissipative viscous processes (time-dependent). Viscoelastic behavior is time dependent while elastic deformations can be time-independent. Plastic materials are somewhat viscous but for simplicity they can be assumed to be time-independent.

Viscoelastic models are composed of different parts including (Ancey, 2005):

Spring: according to Hooke's law, the strain (ε) is proportional to the applied stress (σ) according to:

$$\sigma = E\varepsilon \qquad \text{Eq. (2.8)}$$

with E the elastic modulus. In this law, deformations are time independent and elastic elements represent the possibility of storing energy.

Dashpot: the response of the dashpot, the plunger of which is pushed at velocity $\dot{\varepsilon}$ is described by

$$\sigma = \mu\dot{\varepsilon} \qquad \text{Eq. (2.9)}$$

where (μ) is the viscosity.

The dashpot element represents a dissipative process that occurs as a result of the relative motion between particles. This motion induces friction when there is contact between elements or viscous dampening if there is an interstitial fluid.

14

Viscoelaticity is described by one or more of the above two elements. If the spring and dashpot are mounted in series, the resulting model is called the Maxwell model and it is best suited for viscoelastic fluids. If the spring and dashpot are mounted in parallel, the resulting model is called the Kelvin-Voight model and it is best suited for viscoelastic solids. These models are elementary models and can be combined to get more representative models like the Burgers model. These methods can describe some aspects of the physical behavior, but they still have their own limitations.

2.6.1 Maxwell Model

In this model the dashpot and spring are connected in series, therefore the total deformation is the sum of both deformations:

$$\frac{d\varepsilon}{dt} = \frac{1}{E}\frac{d\sigma}{dt} + \frac{\sigma}{\mu} \qquad\qquad \text{Eq. (2.10)}$$

which has the following solution:

$$\sigma(t) = K e^{-\frac{Et}{\mu}} + \int_{-\infty}^{t} E e^{\frac{E(t'-t)}{\mu}} \dot{\varepsilon}(t')dt' \qquad\qquad \text{Eq. (2.11)}$$

where K is an integration constant, the lower boundary in the integral is arbitrary. K is equal to zero when the stress is finite at time $t=(-\infty)$. The following two cases arise from this equation (Ancey, 2005):

1– for steady state, this equation simplifies to the Newtonian equation $\sigma = \mu\dot{\varepsilon}$.

2– for sudden changes in the stress, the time derivative dominates.

Equation 2.11 can be written as $\sigma(t) = \int_{-\infty}^{t} \left[\frac{\mu}{t_r} e^{\frac{t'-r}{t_r}} \right] \varepsilon'(t')dt' = \int_{-\infty}^{t} \Gamma(t-t')\varepsilon(t')dt'$

where $t_r = \frac{\mu}{E}$ is a relaxation time. The term within the brackets is called the *relaxation modulus*

and the integral takes the form of a convolution product of $\Gamma(t) = \mu \frac{e^{-\frac{t}{t_r}}}{t_r}$ and $\dot{\varepsilon}$.

If the above equation is used in the creep testing it will be reduced to:

15

$$\varepsilon = \sigma\left(\frac{1}{E} + \frac{t}{\mu}\right)$$

Eq. (2.12)

In the Maxwell model, stresses at time t depend on strains at time t and on the strain rate at past time t' but to within a weighting factor that decays exponentially.

2.6.2 Kelvin-Voigt Model

The deformation is described using:

$$\varepsilon = \frac{\sigma}{E}\left(1 - e^{-\frac{t}{t_r}}\right)$$

Eq. (2.13)

where $t_r = \frac{\mu}{E}$ is the relaxation time.

2.6.3 Burgers Model

This model is a combination of the Maxwell model and the Kelvin-Voight models. Deformations are described using

$$\varepsilon = \sigma\left(\frac{1}{E_1} + \frac{1}{E_2}\left(1 - e^{-\frac{t}{t_r}} + \frac{t}{\mu_1}\right)\right)$$

Eq. (2.14)

where $t_r = \frac{\mu_2}{E_2}$ is the relaxation time.

2.6.4 Creep Testing

Creep testing is used to describe the response behavior of solids. In this test, a constant stress is suddenly applied to the material and the strain variation over a range of time is then monitored (Ancey, 2005). Results from the creep test can be used to describe the three distinct responses of the material during the testing. These responses are: (1) the immediate elastic response, (2) delayed elastic response (glassy behavior), where the deformation rate slows with time but becomes steady after long time, (3) the steady state viscous response (when the shear rate of the material is constant so the material is in steady flow).

Visco-elastic models are used to describe material responses during creep testing. Maxwell model is used to describe the immediate elastic response and the steady state response while Burgers model is used to describe the three responses including the glassy response.

2.7 Layered Systems

It is known that the modulus of elasticity of soil and pavement materials is not constant but changes as a function of different factors such as the stress level, moisture content, and temperature (Ullidtz, 1987). On the other hand, the assumption that pavement and subgrade materials are linear does not resemble the actual conditions. Depending on the stress level and strains, the physical response of soil and pavement materials can be categorized as elastic, plastic, viscous, and viscoplastic. Therefore, modeling the pavement and subgrade using only elastic theory might result in inaccurate predictions. This problem can be tackled by assuming nonlinear elastic behavior and by using different techniques to handle the nonlinearity.

2.7.1 Elastic Multilayer Theory

This method was first proposed by Burmister (1943, 1945) as an effort to tackle the limitations of Boussinesq's method. Burmister (1945) simplified the conditions between two layers assuming that all layers are isotropic, elastic, and homogeneous. The top layer, in a two layer system, was assumed to be infinite in extent in the horizontal direction but of finite thickness in the vertical direction. The bottom layer, in a two-layer system, was assumed to be of infinite extent in both horizontal and vertical directions. In addition, Burmister (1945) assumed that the shear and normal stresses outside the limits of the surface loading are equal to zero. Continuity conditions along the interface between layers were considered using two cases. In the first case, full continuity of stress and displacement across the interface was considered assuming full contact between the two layers and a fully activated shear resistance between the layers. In the second case, continuity of normal stresses and normal displacements was only considered by assuming a frictionless interface between the layers.

Vertical stresses based on the two-layer theory are shown in Figure 2.4. As it can be seen from the figure, Burmister's method gave more accurate results than Boussinesq's method since infinite half-space condition is not applicable in pavement and subbase layers because the top

17

layer is always of a finite depth. On the other hand, Burmister's method emphasizes the importance of modeling the interface between the two layers. Boussinesq's method overestimated the vertical stresses at the interface by more than 20% compared to stresses from Burmister's method.

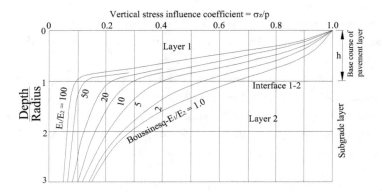

Figure 2.4, Burmister two-layer stress curves (Yoder, 1959).

Actual pavement behavior and conditions are different than those used in the multilayer theory. Pavement materials occupy a finite width and do not have a constant thickness. Spatial variations of material geometry and local discontinuities are not considered when using this method. In addition, idealizing the asphalt and soil material as homogeneous, linear elastic, and isotropic is far from real material behavior. Asphalt has a viscoelastic behavior while granular materials have a non-linear behavior. On the other hand, stresses within the pavement section are not considered in the multilayer theory. Horizontal stresses due to wheel loading and dynamic loads due to road geometry should be considered to correctly simulate the non-linear and viscoelastic responses, respectively (COST 333,1999). Anisotropy of pavement materials due to construction methods and due to the nature of the materials is inevitable and should be considered to account for stress changes in all directions within the pavement section.

Burmister extended his theory to three-layer systems by deriving the settlement equation at the surface of the ground only. Based on Burmister's theory, Acum and Fox (1951) derived a closed form solution for interface stresses under the center of a circular loaded area. In 1962,

18

Schiffman (1962) developed a solution for multilayer elastic systems, which was used by many researchers to develop computer programs for pavement analysis.

2.7.2 Factors Affecting Elasticity

Some of the factors that affect the multilayer elasticity are described below.

2.7.2.1 Anisotropy

Anisotropy has been studied by Van Cauwelaet (1980) assuming a constant ratio between vertical and horizontal moduli. Van Cauwelaet (1980) derived the following equations to incorporate the effect of anisotropy due to a point load. The equations show the solutions below the centerline of a uniformly distributed load.

$$\sigma_z = \frac{szP}{2\pi(1-s)}\left(\frac{1}{\left(\sqrt{s^2z^2+r^2}\right)^3} - \frac{1}{\left(\sqrt{z^2+r^2}\right)^3}\right) \qquad \text{Eq. (2.15)}$$

$$\varepsilon_z = \frac{szP}{2\pi E_V(1-s)}\left(\frac{s^2(\rho+v)}{\left(\sqrt{s^2z^2+r^2}\right)^3} - \frac{1+v}{\left(\sqrt{z^2+r^2}\right)^3}\right) \qquad \text{Eq. (2.16)}$$

$$d_z = \frac{sP}{2\pi E_V(1-s)}\left(\frac{\rho+v}{\sqrt{s^2z^2+r^2}} - \frac{1+v}{\sqrt{z^2+r^2}}\right) \qquad \text{Eq. (2.17)}$$

$$s = \sqrt{\frac{\rho-v^2}{\rho^2-v^2}} \qquad \text{Eq. (2.18)}$$

where E_V is the modulus in compression (vertical), E_h is the modulus in bending (horizontal), v is the Poisson's ratio, and $\rho = E_V / E_h$. For loose to medium dense granular soil with an internal friction angle equal to 30, ρ was found to be 2.25. The definitions of s, z, and r are shown in Figure 2.1.

2.7.2.2 Shear Sensitivity

Shear sensitive material is a material that has an E/G ratio larger than 2(1+v) where E is the Young's modulus, G is the shear modulus, and (v) is the Poisson's ratio. Misra and Sen (1975) proposed the following equations for stresses and displacements in shear sensitive

19

materials with a typical value of 6 (Misra and Sen, 1975) for E/G. The solutions for a point load are:

$$\sigma_z = \frac{zP}{2\pi(\alpha - \beta)} \left(\frac{1}{\left(\sqrt{\beta^2 z^2 + r^2}\right)^3} - \frac{1}{\left(\sqrt{\alpha^2 z^2 + r^2}\right)^3} \right) \qquad \text{Eq. (2.19)}$$

$$\varepsilon_z = \frac{(1+\nu)zP}{2\pi E(\alpha - \beta)} \left(\frac{\alpha^2 A}{\left(\sqrt{\alpha^2 z^2 + r^2}\right)^3} - \frac{\beta^2 B}{\left(\sqrt{\beta^2 z^2 + r^2}\right)^3} \right) \qquad \text{Eq. (2.20)}$$

$$d_z = \frac{(1+\nu)P}{2\pi E(\alpha - \beta)} \left(\frac{A}{\sqrt{\alpha^2 z^2 + r^2}} - \frac{B}{\sqrt{\beta^2 z^2 + r^2}} \right) \qquad \text{Eq. (2.21)}$$

and for a uniform distributed circular load are

$$\sigma_z = \frac{\sigma_o}{\alpha - \beta} \left(\alpha - \beta + \frac{z}{\sqrt{\alpha^2 z^2 + a^2}} - \frac{z}{\sqrt{\beta^2 z^2 + a^2}} \right) \qquad \text{Eq. (2.22)}$$

$$\varepsilon_z = \frac{(1+\nu)\sigma_o}{(\alpha - \beta)E} \left(\beta B \left(\frac{\beta z}{\sqrt{\beta^2 z^2 + a^2}} - 1 \right) - \alpha A \left(\frac{\alpha z}{\sqrt{\alpha^2 z^2 + a^2}} - 1 \right) \right) \qquad \text{Eq. (2.23)}$$

$$d_z = \frac{(1+\nu)\sigma_o}{(\alpha - \beta)E} \left(A \left(\sqrt{\alpha^2 z^2 + a^2} - \alpha z \right) - B \left(\sqrt{\beta^2 z^2 + a^2} - \beta z \right) \right) \qquad \text{Eq. (2.24)}$$

where α^2 and β^2 are the roots of the following equations.

$x^2 + (K' - 2)x + 1 = 0$ where

$$K = \frac{E}{(1+\nu)G} - 1$$

$$K' = \frac{1-K}{1-\nu}$$

$$A = \frac{K - \alpha^2}{\alpha^2 - 1}$$

$$B = \frac{K - \beta^2}{\beta^2 - 1}$$

2.8 In-situ Deflection Measurements

Nondestructive testing is important to measure or describe the material characteristics with minimal damage. In pavement, nondestructive testing plays a major role since cracks and mechanical damage are minimized when pavement cores are reduced. In addition, taking cores and samples from roads that cover many miles can be costly and hence the nondestructive testing provides means by which time and cost can be reduced.

Many systems are available to measure deflections of pavements to back calculate the elastic modulus (E), the foundation support (k or CBR), and the degree of load transfer. Most popular pavement deflection systems include the Benkelman Beam, the Dynaflect, and the Falling Weight Delectometer.

2.8.1 The Benkelman Beam

This is a classical system to measure pavement deformations due to a static load. More details can be found in Bendana et al. (1994).

2.8.2 The Dynaflect System

In this system, sinusoidal vibrations are transmitted to the pavement through steel wheels, and then vertical velocities of the pavement surface at various positions are measured using geophones. Vertical velocities are converted to deflections and the measured deflections (deflection basin) can be estimated. More details can be found in Bendana et al. (1994).

2.8.3 The Falling Weight Deflectometer (FWD)

This system was first introduced 30 years ago in France to test flexible pavements (Ullidtz, 1987). In this system, impact loads (impulse forces) are applied to a loading plate then the vertical velocities of the pavement surface are measured at different locations using geophones. In the LTPP deflection-testing program, geophones are placed at 0, 203, 305, 457, 610, 914, and 1524 mm from the center of the loading plate (Von Quintus and Simpson, 2002). FWD systems, normally, use seven to nine geophones within a distance less than 2 meters with the first sensor placed below the center of the loading plate.

Impulse forces can be varied by adjusting the height of the dropped weights. The pulse load is transferred to the plate through a set of springs and is applied within a short loading time (28 milliseconds). Other values that are measured besides the vertical velocities include the transit time and the duration. Vertical velocities are converted to deflections and the deflection basin can be estimated. Heavy FWD (HFWD) can be used to simulate one wheel of a fully loaded Boeing 747 by producing a maximum instantaneous dynamic force of up to 250 kN in a short loading time between 20 and 25 milliseconds. Figure 2.5 shows a typical configuration for the FWD test.

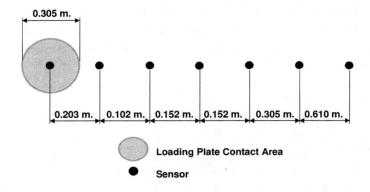

Figure 2.5, FWD typical sensors configuration.

The FWD is believed to generate the most realistic loading function that can be generated by a nondestructive test to simulate actual moving wheel loads (Hoffman, 1983). In addition, Lytton et al. (1986) rated the FWD as the best overall pavement-testing device while Bentsen et al. (1989) commented on the versatility and the time effectiveness of the FWD device.

Pavement responses are measured using a deflection basin which can be created by measuring the responses using uniformly or non-uniformly distributed sensors. More representative basins can be created by placing more sensors around the point where the impulse force is applied. Deflection basin magnitude and shape are dependent on the layer thickness, Poisson's ratio, layer moduli, moduli ratio, and the depth to the stiff layer or the bedrock (Bendana et. al., 1994). Geophones close to the point where the impulse force is applied (inner

22

geophones) are used to measure the composite response of the whole pavement layers. Geophones far from the point where the impulse force is applied (outer geophones) are used to measure the response of the deeper layers. It is a common practice to normalize response data (deflections) to make test results adaptation to other loading conditions easier.

The backcalculated modulus of elasticity, from deflections, is very sensitive to the measured deflections. Calibration of small deflection errors on the order of $1\mu m$-$2\mu m$ will affect the modulus of elasticity significantly (Bendana et. al., 1994).

Stress distribution under pavements is not uniform and controlled by many factors such as the subgrade material, compaction, pavement material, moisture content, and temperature. The actual stress distribution below the pavement can be determined by measuring pavement deflections at different distances from the center of an applied load. Actual stress distributions under pavements are not linear or constant. In granular soils, the soil reaction stresses are higher under the middle of the loading area compared to those at the edge, while in cohesive soils the soil reaction stresses are higher under the edge compared to those under the middle of the loading area (Ullidtz, 1987).

In plate load testing, if the distance from the center of the circular plate is larger than twice the radius of the plate, then, practically, the distributed load can be treated as a point load and hence Boussinesq's theory can be used to estimate the modulus of elasticity based on surface deflections as described in the following equation

$$E = \frac{P(1-v^2)}{\pi d(r,0)}$$
Eq. (2.25)

where $d(r,0)$ is the surface deflection at distance r from the center of the load.

If the modulus of elasticity of the subgrade is not identical at different distances from the applied load, then the assumption that the subgrade is a linear elastic half-space is not valid and either the subgrade is non-linear elastic or it consists of different layers (Ullitdz, 1987). Another reason for non-linearity is the presence of a rigid layer at some depth. Elasticity theory suggests that the deflection at a depth equal to the distance from the load is almost equal to the deflection at the surface (Ullitdz, 1998) according to Eq. (2.25).

23

However, Bodare and Orrje (1985) indicated that the difference between the theoretical dynamic modulus from the FWD and the theoretical static modulus for a homogeneous half-space is minimal when the time to peak load is relatively long. The work of Bush (1980), Roesset and Shao (1985), and Mamlouk (1985) indicated that the difference between the measured dynamic deflections and the predicted static deflections is minimal when a rigid elastic layer is placed at a finite depth below the pavement system.

2.9 Elastic Multilayer Programs

Elastic multilayer programs based on elasticity theory can be divided into forward calculation programs and back-calculation programs. In the forward calculation programs, pavement response is estimated based on known pavement parameters including the modulus of elasticity. In the back-calculation programs, inverse techniques are applied to core forward calculation procedures to estimate the modulus of elasticity based on measured deformations.

2.9.1 Forward Calculation Programs

Different programs have been developed for multilayer elastic analysis including the following:

1- MMOPP: The Mathematical Model of Pavement Performance (MMOPP) was developed by Ullidtz (1987). This program is primarily a research tool and has not yet been used in practice. The program can be used to predict longitudinal roughness, rutting and cracking of a pavement consisting of an asphalt surface course, a granular base layer and subgrade (Danish Road Institute, 2001). Different parameters can be used to study the pavement response including the variation of pavement layer thickness, elastic stiffness, plastic parameters and dynamic load variations along the length of the road.

2-CHEVRON: This program was written based on the work of Michelow (1963). The program capabilities include computing vertical, tangential, radial, and shear stresses, and vertical and radial strains. Program's limitations included the incapability of computing displacements and principal stresses. The program was limited to 5 layers in its early stages then was expanded to fifteen (15) layers in 1967 (Ullitdz, 1987). The CHEVRON model was used by the Asphalt

24

Institute in 1991 to develop the Mechanistic-Empirical Thickness Design Manual (Asphalt Institute, 1991).

3-ELSYMS: This program was built based on CHEVRON with more capabilities. These capabilities included the ability to compute displacements and principal stresses and the ability to handle multi-wheel loads.

4-BISTRO: This program was written based on the work of Schiffman (1962). The program capabilities included computing stresses, strains, and displacements, assuming rough interfaces between layers. The program can account for multi-wheel loads.

5-BISAR: This program was built based on BISTRO with more capabilities. The program can consider the tangentional surface loadings and include layer interface. This program was developed by Shell International Petroleum Company Limited and used in their Pavement Design Manual (Shell, 1978).

6-MultiSmart3D: This program was developed by the Computer Modeling and Simulation Group at the University of Akron (Pan and Alkasawneh, 2006). This program is based on the innovative computational and mathematical techniques for multilayered elastic systems (e.g., Pan, 1989a,b, 1990, 1997). The MultiSmart3D program uses the propagator matrix method to solve only two 2×2 systems of linearly algebraic equations no matter how many layers are in the pavement system. In addition, adaptive Gauss quadrature is used along with an acceleration approach for fast and accurate calculation of integrations. The program is capable of calculating the pavement responses at any arbitrary number of points at any location in the pavement system due to any arbitrary loading configuration (uniform or nonuniform loading) and shape (regular or irregular shapes). Moreover, the program can be used for any arbitrary number of layers with any thickness.

2.10 Back Calculation of Elastic Moduli

Elastic modulus is an important property of pavement materials. Different methods have been proposed by researchers to estimate the elastic modulus based on laboratory bending tests and empirical equations (Bonnaure et al., 1977), wave propagation methods (Avramesco, 1968; Heukelom and Foster, 1960; Jones, 1963, 1965; Jones and Thrower, 1965; Szendrei and Freeme, 1970; Cho and Lin, 2001), and the Falling Weight Deflectometer (FWD).

Elastic modulus cannot be determined accurately based on empirical equations since actual field conditions, loading conditions, and traffic conditions vary along the same section. In addition, some of the input parameters in the empirical equations cannot be determined for an existing pavement while it can be determined during the mix design stage of the pavement. For practical purposes, elastic moduli need to be determined on site to measure the interaction and effect of different factors.

Wave propagation methods are non-destructive insitu testing methods used to determine the elastic modulus and the thickness of the pavement. These methods are based on measuring the phase difference between the vibration source that transmits sinusoidal stress waves of a frequency set into the test object, and the vibration of the receivers, which are placed at different distances from the source (Martincek, 1994). The vibration sources, usually, have small excitation energy which makes the test non-destructive as the resulting dynamic stresses are low and cannot affect the state of the pavement.

2.10.1 Classical Backcalculation Methods

Methods for the backcalculation of the pavement moduli using data from nondestructive testing can be grouped into three general methods (Anderson, 1988): (1) simplified methods, (2) gradient relaxation methods, and (3) direct interpolation methods. These methods rely on some assumptions and simplifications that should be made to facilitate the backcalculation process. For example, the thickness of each layer is assumed to be constant (uniform) and normally is obtained based on the construction records or pavement coring. The main steps of the backcalculation procedure consists of:

1-Define the input parameters of the pavement system including: thickness of each layer, Poisson's ratio, …etc.

2-Assume moduli seed values for the pavement system. Seed moduli values can be assumed based on experience or based on typical moduli values. Moduli values can be different based on the forward method implemented in the backcalculation program.

3-Calculate the pavement deflections, using the forward program, at the FWD geophone locations (along the surface).

26

4-Compare the calculated deflections with the measured deflections. If the difference between the calculated and measured deflections is acceptable, then the assumed layer moduli are the actual moduli. Otherwise, the assumed layer moduli are not the actual moduli and the assumed moduli should be refined.

5-Repeat steps 2 through 4 if necessary.

2.10.1.1 Simplified Methods

The simplified methods are backcalculation methods that depend on the equivalent thickness and the deflection geometry methods as the forward calculation methods. Ullidtz (1973) developed a backcalculation computer program named ELMOD based on the equivalent thickness method while Van der Loo (1982) developed another program using the method of surface curvature to backcalculate the layer moduli. These programs have the same drawbacks that are associated with their forward calculation methods. Therefore, both programs are not widely used due to reliability issues.

2.10.1.2 Gradient Relaxation Methods

The gradient relaxation methods are the most common method for the backcalculation of pavement moduli. The method was first proposed by Michelow (1963). Other programs developed later including BISDEF, ELSDEF, and CHEVDEF (Bush, 1980; Bush and Alexander, 1985), which use BISAR, ELSYM5, and CHEVRON, respectively, as their forward calculation programs.

The main idea is to form gradient matrices then to solve these matrices for solutions that give the minimal difference between calculated and measured deflections. The relation between the predicted deflection Δ_j at sensor j and the layer moduli (E) for a pavement system of n layers is given by (Anderson, 1988):

$$\Delta_j = f(E_1, E_2, E_3, ..., E_n) \qquad \text{Eq. (2.26)}$$

The deviation in deflection at sensor j between the measured (D_j) and the predicted deflection is:

$$\delta_j = D_j - \Delta_j = f(E_1, E_2, E_3, ..., E_n) \qquad \text{Eq. (2.27)}$$

The sum of the squares of all sensors' deviations is:

27

$$\sum_{j=1}^{m} \delta_j^{\,2} = \sum_{j=1}^{m} [D_j - f(E_1, E_2, E_3, ..., E_n)] \qquad \text{Eq. (2.28)}$$

where m is the number of sensors (deflection points).

The above equation, then, is derived with respect to the unknown moduli (partial derivatives) to form a solution matrix of size n equation. Then, the partial derivatives are approximated numerically by forming gradient equations. The gradient equations are formed by calculating the deflections of an initial set of moduli (E^o). Another set of moduli values (E^1) is assumed where the moduli values are all the same as in set (E^o) but with one of the moduli changed. Then, the deflections of the second set are calculated. The deflections at sensor j as a function of the gradient equation can be estimated using:

$$\Delta_j = A_{ji} + S_{ji} \log_{10}(E_i) \qquad \text{Eq. (2.29)}$$

where E_i is the unknown modulus of layer i, $S_{ji} = \dfrac{\Delta_i^{\,o} - \Delta_j^{\,1}}{\log_{10} E_i^{\,o} - \log_{10} E_i^{\,1}}$, $A_{ji} = \Delta_i^{\,o} - S_{ji} E_i^{\,1}$, $E_i^{\,o}$ is

the assumed modulus for layer i in the initial set of moduli, $E_i^{\,1}$ is the assumed modulus for layer i in the second set of moduli, $\Delta_i^{\,o}$ is the calculated deflection at sensor j for $E_i^{\,o}$, $\Delta_i^{\,1}$ is the calculated deflection at sensor j for $E_i^{\,1}$.

The deflection at sensor j as a function of all unknown moduli values E_i:

$$\Delta_j = \Delta_j^{\,o} + \sum_{i=1}^{n} S_{ji} \left(\log_{10} E_i - \log_{10} E_i^{\,o} \right) \qquad \text{Eq. (2.30)}$$

Then, $\Delta_j^{\,o}$ can be estimated using one of the unknown moduli (e.g., modulus of layer number 3):

$$\Delta_j^{\,o} = A_{j3} + S_{j3} \log_{10} E_3^{\,o} \qquad \text{Eq. (2.31)}$$

Based on Eqs. (2.29) and Eq. (2.30), Eq. (2.31) can be written as:

$$\Delta_j = A_{j3} + S_{j3} \log_{10} E_3^{\,o} + \sum_{i=1}^{n} S_{ji} \left(\log_{10} E_i - \log_{10} E_i^{\,o} \right) \qquad \text{Eq. (2.32)}$$

Then, Eq. (1.29) becomes:

$$\sum_{j=1}^{n} \delta_j^{\,2} = \sum_{j=1}^{m} \left[D_j - A_{j3}^{\,o} - S_{j3} \log_{10} E_3^{\,o} - \sum_{i=1}^{n} S_{ji} \left(\log_{10} E_i - \log_{10} E_i^{\,o} \right) \right]^2 \qquad \text{Eq. (2.33)}$$

28

The squared errors in deviation in the above equation can be minimized by taking the partial derivative with respect to each of the unknown moduli, then setting the partial derivatives equal to zero.

Then, the following matrix should be solved:

$$[B]\{E\}=\{C\} \qquad \text{Eq. (2.34)}$$

where:

$$C_k = \sum_{j=1}^{m} S_{jk} \left[D_j - A_{j3} - S_{j3} \log_{10} E_3 - \sum_{i=1}^{3} S_{ji} \log_{10} E_i \right] \qquad \text{Eq. (2.35)}$$

$$B_{ki} = \sum_{j=1}^{m} S_{jk} S_{ji} \qquad \text{Eq. (2.36)}$$

where i, k are the layer numbers, and E_k is the unknown layer moduli associated with equations (2.34) through (2.36).

The above procedure is dependent on the forward calculation method. Increasing the number of layers will increase the time spent searching for the solution.

2.10.1.3 Direct Interpolation Methods

This method was developed by Uzan et al. (1988) based on the computer program MODULUS (Uzan, 1986). In this method, the user selects the structural response model and then the moduli are backcalculated using direct interpolation from a set of solutions. The method requires the existence of a table of pre-calculated responses for a wide range of moduli combinations. Therefore, the accuracy of the program can be limited since the accuracy of the interpolation will be dependent on the number of data points and the number of cases considered to create the table.

2.10.2 Elastic Multilayer Backcalculation Programs

Most backcalculation programs apply iterative and inverse techniques on the multilayer elastic theory to estimate the modulus of elasticity. As a minimum, back calculation methods using multilayer elastic theory require the thickness of the layer and the Poisson's ratio to be

29

known. Typical values of the Poisson's ratio and the modulus of elasticity for different pavement materials are given in Table 2.2 (AASHTO, 1993) and Table 2.3, respectively. Guidelines for backcalculating the elastic properties of pavement from the FWD test are described in ASTM D5858.

Table 2.2, Typical Poisson's ratios of pavement materials,
modified after (AASHTO, 1993).

Material	Range	Remarks	Typical Value
Portland Cement Concrete	0.10-0.20		0.15
Hot Mix Asphalt/ Asphalt Treated Base	0.15-0.45	Depends on temperature. For temperatures less than 30°C use 0.15; for temperatures higher than 50°C use 0.45.	0.35
Cement Stabilized Base	0.15-0.30	When sound free of cracks use 0.15; with cracks use 0.30	0.20
Granular Base/Subbase	0.30-0.40	For crushed material use low values; for unprocessed rounded gravel/sands use high values	0.35
Subgrade Soils	0.30-0.50	For cohesionless soils use values near 0.30; for very plastic/cohesive clays use 0.50.	0.40

Table 2.3, Typical values of modulus of elasticity for pavement materials, after (AASHTO, 1993).

Material	Range (MPa)	Typical Value (MPa)
Hot-Mix Asphalt	1,500-3,500	3,000
Portland Cement Concrete	20,000-55,000	30,000
Asphalt-Treated Base	500-3,000	1,000
Cement-Treated Base	3,500-7,000	5,000
Lean Concrete	7,000-20,000	10,000
Granular Base	100-350	200
Granular Subgrade Soil	50-150	100
Fine-Grained Subgrade Soil	20-50	30

Methods for back calculating the elastic modulus of a layered pavement system can be divided based on the inverse approach, as the following (von Quintus and Simpson, 2002):

1-The equivalent thickness method (e.g. ELMOD and BOUSDEF programs).

2-The optimization method (e.g. MODULUS and WESDEF programs).

3-The iterative method (e.g. MODCOMP and EVERCALC programs).

The following is a brief description of some of the available back calculation programs:

1- MODCOMP 3: This program is used to back calculate the modulus of elasticity of pavement layers based on the work of Irwin (Irwin, 1983, 1992). The modulus of elasticity is back calculated using iterative methods where the modulus is adjusted until it matches the measured deflections from the geophones (sensors). The program can handle data from up to ten deflection sensors at up to six load levels. The maximum number of layers with unknown moduli can be up to twelve (12) layers with a recommendation of using five or six unknown moduli only.

2-WES5: This program is a linear elastic program that was developed by the Corps of Engineers. This program uses algorithms that make the calculations faster than many other programs and has no copy right restrictions.

3-MODULUS: This program was developed in 1986 (Uzan, 1986) based on the Corps of Engineers WES5 program. The program can handle up to five layers and can be used to estimate the optimal number of sensors to be used in the backcalculation process. In addition, the program can, automatically, estimate the depth of the stiff layer (rock) and the weighted factors for each layer. The program uses Hooke-Jeeves pattern search algorithm which provides a definite convergence for the backcalculated values.

4-BISDEF: This program is based on the calculation process in the BISAR program (Bush, 1985). The program uses the iterative gradient relaxation method to backcalculate the moduli. Sensitivity analysis showed that the program is sensitive to the initial moduli values (seed moduli) and in many cases did not converge (Anderson, 1988).

5-ELSEDEF: This program is based on the calculation process in ELSYM5 (Lytton et. *al.*, 1990).

6-CHEVDEF: This program is based on the calculation process in CHEVRON (Warren and Diekmann, 1963).

7-COMDEF: This program was developed in 1989 (Anderson, 1989) to back calculate the modulus of elasticity using a stored database of deflection basins generated with CHEVRON. In this method, a set of pavement system parameters is selected, including the layer thickness and the modulus of elasticity for each layer, and the deflection basin is calculated for that set and stored in a database. Then the measured data from the in-situ deflection testing are compared with those in the stored database to estimate the required values. However, the program did not work well with cases that involved thick layers or very shallow deflection basins (small deflections) making the program inappropriate to some of the airfield pavements due to the limitations of the database.

8-ILLI-BACK: This program was developed to back calculate the slab modulus and the subgrade reaction modulus for a two-layer rigid pavement system (Ioannides et. *al.*, 1989). The program uses the area of the deflection basin and the maximum deflection as inputs to Westergaard's deflection equation assuming a unique relationship between the area and the radius of relative stiffness of the pavement system.

2.10.3 Limitations of The Backcalculation Programs

Back calculation programs suffer from different limitations based on the assumptions and simplifications used to develop the program which reflects, in most cases, the limitations of the core multilayer elastic programs. One way to overcome these limitations was proposed previously (Von Quintus and Killingsworth, 1997) using correction factors based on comparisons between laboratory-measured and back-calculated elasticity moduli. However, these correction factors were dependent on the backcalculation program and hence should be used carefully.

Some of the limitations observed in the backcalculation process carried out in different studies (Von Quintus and Simpson, 2002; George, 2003) are listed below.

1-Many programs are limited by the number and the thickness of layers used in the analyzed pavement system.

2-Many programs assume a linear elastic behavior with few considering the non-linear behavior. Unfortunately, programs that account for viscoelastic (time-dependent) or elastoplastic (inelastic) behavior of materials have not been successful in estimating consistent and reliable results.

3-Actual modulus of elasticity of the layer is not calculated because many factors affecting the modulus cannot be included in the current used methods. These factors include stress sensitivity, discontinuities and anomalies such as variations in layer thickness, localized segregation, cracks, and the combination of similar materials into a single layer. Therefore, the back calculated modulus is known as the "effective" or "equivalent" modulus.

4-Most programs use static back calculations whereas the applied load in the FWD is a dynamic load. Standardized back calculation procedures are available only for the static case (ASTM D5858) while the dynamic case is more complex.

5-A large number of iterations is required, if the model is based on the linear elastic assumption, at varying load levels (different drop heights) to identify the stress sensitivity of unbound pavement materials and soils. In other words, forward analysis (deflection calculation) of pavement systems using linear response models is relatively fast while back calculation using linear models to estimate the nonlinear properties require relatively large number of iterations.

6-Most programs are developed based on quasi-nonlinear response models. This indicates that the modulus of elasticity for a layer is the same in the horizontal as well as in the vertical

33

directions even if the state of the stress within the layer is changing. Variation of the modulus of elasticity within the pavement layer has not been considered. In addition, the modulus can vary by changing the load level.

7-Solution convergence is not guaranteed for many programs. The starting modulus (the seed) should be adjusted based on the user's experience and judgment to arrive at a reasonable modulus that fit the elastic modulus range for the layer and the desired root mean square error (RMSE). This makes the backcalculation process highly dependent on the user rather than completely automated.

8-Accuracy of the back calculated properties using elastic models vary with depth. Accuracy of the back-calculated results may vary between 1 percent to 20 percent with the high value for the surface layer and the low value for the subgrade layer.

9-Most programs are unable to model plastic flow or lateral movement of the underlying materials close to the surface. This limitation can result in the back calculation of unrealistically high elastic moduli.

10- Modulus of elasticity is stress dependent in coarse and fine-grained unbound materials. Hot mix asphalt (HMA) layers are assumed to be linear elastic and therefore they are stress independent. However, backcalculation programs show stress sensitivity (stress dependent elasticity modulus) that is dependent on the drop height in the FWD test. This contradiction in behavior is attributed to limitations in the backcalculation process which ignores the interface condition near the surface and the inability to simulate any damage in the bound surface layers rather than the true stress sensitivity of the layer.

11-Subdividing the subgrade into more than two layers can significantly improve the matching between the measured and calculated deflection basins resulting in low root mean square errors. However, most programs have limitations regarding the maximum number of layers.

12-All back calculation programs are sensitive to the rigid layer location below the subgrade/subbase interface. However, the SHRP recommended (SHRP-P-655, 1993) placing the rigid layer at a depth of 50 feet below the surface if no information is available.

13- The backcalculated moduli is not unique. Different backcalculated moduli can be obtained for the same root mean square error.

2.10.4 Backcalculated Resilient Modulus versus Lab Testing

The difference between backcalculated subgrade resilient modulus from the FWD and laboratory resilient modulus has been studied by many researcher (Daleiden et *al.*, 1994; Akram et al., 1994; Nazarian et *al.*, 1995; Von Quintus and Killingworth, 1997; Seeds et *al.*, 2000; George, 2003).

It was found by Daleiden et *al.* (1994), Akram et al.(1994), Seeds et *al.* (2000) and Nazarian et *al.*(1995) that there is no unique relation between the backcalculated resilient modulus and the laboratory measured resilient modulus. The ASSHTO Guide (1993) suggested that the backcalculated modulus is three times the laboratory modulus, while Von Quintus and Killingworth (1997) suggested using factors, to match the backcalculated and the laboratory moduli, developed using a multi layer elastic program. However, the suggested correction factors are highly dependent on the backcalculation program and should be used carefully.

Besides the spatial variations and nonlinearity of materials that contribute to this contradiction, the FWD test measures deflections due to stresses in 3D directions while deflections in conventional multilayer back calculation programs are assumed to be in 2D instead. Conventional laboratory testing procedures estimate the modulus based on measurements in the vertical direction only. Programs that consider the 3D effect and the variation of the resilient modulus in both the horizontal directions (anisotropy) should be developed to overcome such problems.

Variation of the modulus of elasticity in subgrade material can be observed even if the subgrade contains one type of material. This variation can be observed when the moisture content of the material varies with depth within the subgrade (George, 2003). In the backcalculation as well as the forward calculation procedures, the assumption that the subgrade layer has a constant modulus with depth, is based on the need for simplification since most programs have limitations on the number of layers.

Another reason for anisotropy in subgrade material is compaction. It is a common practice to prepare the subgrade material by compacting the subgrade lifts during construction, which can be from 6 to 12 inches based on the used equipment and material. Compaction energy applied to each layer will change the stress distribution and hence will change the modulus of elasticity in both directions resulting in an anisotropic state.

2.11 Conclusions

Elasticity theory has been used for many years by engineers to analyze pavement responses due to traffic loading due to its simplicity and cost effectiveness. In addition, field and lab test data showed that the elasticity theory can provide accurate results and can be practical more than other methods since less controlling parameters are needed. The elasticity theory was further implemented in computer programs to facilitate the analysis of pavement systems. These computer programs are limited to a certain number of pavement layers due to the limitations associated with the mathematical formulation of the closed form solutions. The maximum number of layers that can be used in pavement programs is 20 layers. This limited number of layers limits the modeling of pavement systems where temperature variation with depth is observed. Therefore, the need for a more robust and flexible forward calculation (calculates response form current known moduli) program should be a priority.

However, the backcalculation of the pavement layer moduli, using the available pavement programs and backcalculation techniques, showed many drawbacks. These drawbacks include the difficulty of finding the optimal solution and the possibility of having more than one set of moduli for the same set of pavement deflections. This difficulty stems from the fact that seed moduli are used in the backcalculation process leading to a local solution rather than a global solution for the problem. In addition, most backcalculation techniques are not computationally effective or robust leading to divergence (no solution is obtained) in some cases. Therefore, the development of a robust backcalculation program (calculates moduli from known deflections) that can be used for a wide spectrum of pavement loading and modeling conditions is vital.

CHAPTER III

EFFECT OF THE ROOT-MEAN-SQUARE ERROR

ON PAVEMENT RESPONSE

3.1 Introduction

The elastic modulus is an important property of pavement materials. Different methods have been proposed by researchers to estimate the elastic modulus based on laboratory bending tests and empirical equations (Bonnaure et al., 1977), wave propagation methods (Cho and Lin, 2001), and the Falling Weight Deflectometer (FWD).

The elastic modulus in pavements cannot be determined accurately based on empirical equations since actual field conditions, loading conditions, and traffic conditions vary. In addition, some of the input parameters in the empirical equations cannot be determined for an existing pavement. Therefore, for practical purposes, elastic moduli need to be determined using in-situ methods to include the effect of different field factors.

The FWD test is currently the most widely used nondestructive method in pavement engineering. The test involves applying impact loads (impulse forces) to a loading plate and measuring the vertical displacement of the pavement surface at different locations using velocity sensors. The FWD system normally uses seven to nine geophones within a distance less than 2 meters with the first sensor below the center of the loading plate. The measured deflections from the FWD test along the pavement surface are then utilized to backcalculate the modulus of elasticity in each layer. This method, however, suffers from different limitations since backcalculating the modulus of elasticity does not always ensure an accurate estimate of the modulus where a seed modulus is required for each layer in the backcalculation procedure. Therefore, the backcalculation of the elastic moduli does not provide a unique solution and in many cases is user-dependent. It is common to assess the accuracy of the backcalculated elasticity moduli by assessing the accuracy between the measured deflections from the FWD and the calculated deflections using the backcalculated set of elastic moduli.

In this book we investigate the current practice of rejecting and accepting the backcalculated elasticity moduli based on a tolerance value between the measured and calculated deflections.

3.2 Error Tolerance in The Backcalculation Procedure

The backcalculation of elasticity moduli is commonly carried out by assuming a set of pavement-layer moduli (seed moduli) that can produce a deflection basin similar to the measured one from the FWD test. In order to minimize the error between the measured and calculated deflections, the relative root-mean-square error (RMSE) is used to control the convergence of the backcalculated deflections and to assess the acceptance and rejection of the final set of pavement moduli. The RMSE is computed by:

$$RMSE = \sqrt{\frac{1}{n} \sum_{i=1}^{n} \left(\frac{d_i - D_i}{D_i} \right)^2} \times 100\% \qquad \text{Eq. (3.1)}$$

where, *RMSE* is the relative root-mean-square error, n is the total number of the deflection measurement points, d_i is the backcalculated deflection at point i, and D_i is the measured deflection at point i. When the RMSE value decreases, the accuracy of the backcalculated elasticity moduli is assumed to increase as the error between the measured and calculated deflections decreases.

In the Long Term Pavement performance (LTPP) test sections, an RMSE of 3% was used as an acceptable error (Von Quintus and Simpson, 2002). In addition, Von Quintus and Simpson (2002) showed that the selection of a 2% RMSE does not necessarily result in convergence in the backcalculated elasticity moduli in all cases. In general, their results indicated (Von Quintus and Simpson, 2002) that RMSE values less than 3% have little effect on the average backcalculated elastic moduli. In practice, RMSE values larger than 1% can be achieved quickly (Harichandran et al., 1994). Therefore, the most commonly used value for the RMSE is between 1% and 3%. However, it is believed that achieving lower RMSE will always enhance the backcalculated elastic moduli and therefore more accurate results can be obtained.

3.3 Backcalculation Study

The majority of the available research investigates the difference between the RMSE values from different backcalculation programs (Fwa et al. 1997), the effect of other factors on the quality of the deflection data (Mehta and Roque 2003), the effect of the seed generation on the RMSE values (Fwa and Rani, 2005), and the effect of other factors on the FWD data. To the best of the authors' knowledge, however, the effect of the backcalculated elastic moduli and the associated RMSE on the strain and stress responses in flexible pavements has not been discussed so far.

In this chapter, to study the effect of the RMSE on the strains and stresses in flexible pavements, a three-layer pavement section was selected. The flexible pavement section and the backcalculated elastic moduli were reported by Anderson (1988) and were shown in Tables 3.1 and 3.2. In this book, the responses and the associated error at a total of 11 points (as shown in Tables 3.3 and 3.4) were calculated for each case: responses along the ground surface at common locations of the velocity sensors used in the FWD test along the pavement profile (points 1 through 7), at the middle of the AC layer (point 8), at the bottom of the AC layer (point 9), at the middle of the base layer (point 10), and at the top of the subgrade (point 11). For all cases, responses were calculated for a circular load with a radius of 150 mm and a pressure of 690 kPa. The Poisson's ratio for all layers was equal to 0.35.

The pavement response was calculated using the *MultiSmart3D* program. The *MultiSmart3D* program is a fast and accurate software tool developed by the Computer Modeling and Simulation Group at the University of Akron, and it is based on the innovative computational and mathematical techniques for multilayered elastic systems (Pan 1989a,b; Pan 1990; Pan 1997). The program is capable of analyzing any pavement system regardless of the number of layers, the thickness of each layer, the number of response points, and the shape of the applied pressure at the surface of the pavement.

The RMSE for each case was calculated using Eq. 3.1 and shown in Table 3.3. In addition, Table 3.3 shows the relative errors using the exact and backcalculated elastic moduli. The relative error is defined as:

$$RE = \left| \frac{Exact - Calculated}{Exact} \right| x100\% \qquad \text{Eq. (3.2)}$$

This definition of the relative error is also used to compare the calculated strains and stresses based on the exact and backcalculated models. In other words, *Calculated* is the response using the backcalculated elastic moduli, and *Exact* is the response using the exact elastic moduli.

Pavement responses from the exact and backcalculated elastic moduli (Cases 1 through 5) are shown in Tables 3.5 through 3.10. In addition, Tables 3.5 through 3.10 show the relative errors associated with each response.

Table 3.1, Parameters of the Original (Exact) Layers.

Case	AC Thickness (mm)	Base Thickness (mm)	Exact Modulus (MPa)		
			AC	Base	Subgrade
Case 1	381.0	152.4			
Case 2	381.0	152.4			
Case 3	381.0	152.4	3447.3785	13789.5140	68.9476
Case 4	381.0	152.4			
Case 5	381.0	152.4			

Table 3.2, Parameters of the Backcalculated Layers.

Case	Backcalculated Modulus (MPa)		
	AC	Base	Subgrade
Case 1	3506.1839	13507.0427	69.0096
Case 2	3359.2980	15717.2743	68.6304
Case 3	3656.5723	12088.8670	69.4302
Case 4	3410.8018	13371.5676	68.7063
Case 5	3657.1101	12017.7062	69.6715

Table 3.3, Relative Errors in Moduli and RMSEs using the Exact (Table 3.1) and Backcalculated (Table 3.2) Moduli.

Case	AC (%)	Base (%)	Subgrade (%)	RMSE (%)
1	1.71	2.05	0.09	0.24
2	2.55	13.98	0.46	0.22
3	6.07	12.33	0.70	0.65
4	1.06	3.03	0.35	0.69
5	6.08	12.85	1.05	0.85

Table 3.4, Coordinates and Locations of the Response Points.

41

Point	x (mm)	y (mm)	z (mm)	Location
1	0.0	0.0	0.0	AC Surface
2	304.8	0.0	0.0	
3	609.6	0.0	0.0	
4	914.4	0.0	0.0	
5	1219.2	0.0	0.0	
6	1524.0	0.0	0.0	
7	1828.8	0.0	0.0	
8	0.0	0.0	190.5	Middle of AC Layer
9	0.0	0.0	380.9	Bottom of AC Layer
10	0.0	0.0	457.2	Middle of Base Layer
11	0.0	0.0	533.5	Top of Subgrade Layer

Table 3.5, Exact Responses from the Original (Exact) Pavement Profile.

Point	ε_x	ε_y	ε_z	σ_x	σ_y	σ_z	u_z
	(μm/m)	(μm/m)	(μm/m)	(kPa)	(kPa)	(kPa)	(mm)
1	-57.77	-57.7	-62.49	-677.95	-677.95	-690.00	0.21
2	-7.00	-27.08	18.35	-64.75	-116.02	0.00	0.17
3	-8.44	-18.22	14.35	-58.21	-83.17	0.00	0.16
4	-3.32	-14.08	9.37	-32.39	-59.87	0.00	0.15
5	-0.04	-10.94	5.91	-15.20	-43.02	0.00	0.13
6	1.74	-8.56	3.67	-4.93	-31.24	0.00	0.12
7	2.70	-6.76	2.18	1.32	-22.83	0.00	0.11
8	22.57	22.57	-88.36	-71.10	-71.10	-354.37	0.19
9	1.60	1.59	-21.33	-49.96	-49.97	-108.51	0.18
10	10.32	10.32	-12.91	197.33	197.33	-39.95	0.18
11	21.43	21.43	-85.56	-1.45	-1.45	-6.91	0.18

Table 3.6, Response (Top Row) and Relative Error (Bottom Row in Parenthesis) Using Case 1 Backcalculated Layer Model.

Point	ε_x (μm/m)	ε_y (μm/m)	ε_z (μm/m)	σ_x (kPa)	σ_y (kPa)	σ_z (kPa)	u_z (mm)
1	-57.10 (1.173)	-57.10 (1.173)	-61.13 (2.179)	-679.52 (0.232)	-679.52 (0.232)	-690.00 (0.000)	0.21 (0.465)
2	-7.04 (0.592)	-26.87 (0.788)	18.26 (0.505)	-65.72 (1.501)	-117.21 (1.021)	0.00 (0.000)	0.17 (0.223)
3	-8.34 (1.233)	-18.07 (0.777)	14.22 (0.922)	-58.59 (0.651)	-83.88 (0.851)	0.00 (0.000)	0.16 (0.197)
4	-3.27 (1.408)	-13.96 (0.854)	9.28 (0.960)	-32.59 (0.610)	-60.34 (0.794)	0.00 (0.000)	0.15 (0.186)
5	-0.04 (7.515)	-10.84 (0.876)	5.86 (0.901)	-15.32 (0.742)	-43.37 (0.806)	0.00 (0.000)	0.13 (0.171)
6	1.72 (1.039)	-8.49 (0.875)	3.64 (0.833)	-4.99 (1.047)	-31.50 (0.829)	0.00 (0.000)	0.12 (0.158)
7	2.67 (1.054)	-6.70 (0.865)	2.17 (0.739)	1.31 (0.722)	-23.03 (0.858)	0.00 (0.000)	0.11 (0.146)
8	22.16 (1.823)	22.16 (1.823)	-86.76 (1.808)	-71.03 (0.099)	-71.03 (0.099)	-353.91 (0.129)	0.19 (0.317)
9	**1.91** **(19.686)**	**1.91** **(19.686)**	-21.17 (0.742)	-47.60 (4.722)	-47.60 (4.722)	-107.55 (0.879)	0.18 (0.241)
10	10.51 (1.910)	10.51 (1.910)	-13.14 (1.777)	197.19 (0.074)	197.19 (0.074)	-39.50 (1.118)	0.18 (0.255)
11	21.56 (0.611)	21.56 (0.611)	-85.49 (0.089)	-1.42 (1.720)	-1.42 (1.720)	-6.90 (0.248)	0.18 (0.264)

44

Table 3.7, Response (Top Row) and Relative Error (Bottom Row in Parenthesis) Using Case 2 Backcalculated Layer Model.

Point	ε_x (μm/m)	ε_y (μm/m)	ε_z (μm/m)	σ_x (kPa)	σ_y (kPa)	σ_z (kPa)	u_z (mm)
1	-58.31 (0.930)	-58.31 (0.930)	-65.18 (4.306)	-672.90 (0.745)	-672.90 (0.745)	-690.00 (0.000)	0.21 (0.035)
2	-6.75 (3.632)	-27.02 (0.227)	18.18 (0.926)	-62.04 (4.186)	-112.48 (3.051)	0.00 (0.000)	0.17 (0.407)
3	-8.70 (3.091)	-18.23 (0.088)	14.50 (1.039)	-57.74 (0.802)	-81.45 (2.061)	0.00 (0.000)	0.16 (0.307)
4	-3.52 (6.211)	-14.17 (0.660)	9.53 (1.719)	-32.48 (0.265)	-58.97 (1.500)	0.00 (0.000)	0.15 (0.175)
5	**-0.17** **(300.543)**	-11.05 (1.009)	6.04 (2.154)	-15.45 (1.594)	-42.52 (1.180)	0.00 (0.000)	0.13 (0.052)
6	1.66 (4.465)	-8.67 (1.265)	3.77 (2.728)	-5.25 (6.419)	-30.96 (0.894)	0.00 (0.000)	0.12 (0.066)
7	2.66 (1.604)	-6.86 (1.484)	2.26 (3.539)	**0.98** **(25.332)**	-22.69 (0.620)	0.00 (0.000)	0.11 (0.180)
8	23.21 (2.829)	23.21 (2.829)	-91.06 (3.060)	-71.82 (1.021)	-71.82 (1.021)	-356.17 (0.509)	0.19 (0.302)
9	**0.30** **(81.365)**	**0.30** **(81.365)**	-21.15 (0.831)	**-58.94** **(17.970)**	**-58.94** **(17.970)**	-112.31 (3.509)	0.18 (0.468)
10	**9.17** **(11.155)**	**9.17** **(11.155)**	**-11.52** **(10.792)**	199.20 (0.945)	199.20 (0.945)	-41.63 (4.217)	0.18 (0.405)
11	20.15 (5.994)	20.15 (5.994)	-83.18 (2.781)	-1.52 (4.868)	-1.52 (4.868)	-6.77 (2.037)	0.18 (0.350)

Table 3.8, Response (Top Row) and Relative Error (Bottom Row in Parenthesis) Using Case 3 Backcalculated Layer Model.

Point	ε_x (μm/m)	ε_y (μm/m)	ε_z (μm/m)	σ_x (kPa)	σ_y (kPa)	σ_z (kPa)	u_z (mm)
1	-55.76	-55.76	-57.53	-685.22	-685.22	-690.00	0.21
	(3.484)	(3.484)	(7.948)	(1.072)	(1.072)	(0.000)	(1.228)
2	-7.24	-26.58	18.21	-68.93	-121.30	0.00	0.17
	(3.401)	(1.870)	(0.787)	(6.460)	(4.548)	(0.000)	(0.380)
3	-8.02	-17.84	13.92	-59.42	-86.03	0.00	0.16
	(5.037)	(2.058)	(3.001)	(2.085)	(3.444)	(0.000)	(0.391)
4	-3.07	-13.71	9.04	-32.78	-61.61	0.00	0.15
	(7.473)	(2.602)	(3.532)	(1.229)	(2.914)	(0.000)	(0.457)
5	**-0.05**	-10.62	5.91/5.69	-15.28	-44.19	0.00	0.13
	(219.214)	(2.883)	(3.709)	(0.522)	(2.703)	(0.000)	(0.513)
6	1.75	-8.30	3.53	-4.81	-32.04	0.00	0.12
	(0.508)	(3.054)	(3.963)	(2.409)	(2.539)	(0.000)	(0.568)
7	2.66	-6.54	2.09	**1.55**	-23.38	0.00	0.11
	(1.470)	(3.182)	(4.322)	**(17.314)**	(2.398)	(0.000)	(0.621)
8	21.21	21.21	-82.83	-70.20	-70.20	-352.01	0.19
	(6.017)	(6.017)	(6.256)	(1.260)	(1.260)	(0.665)	(0.669)
9	**3.29**	**3.29**	-21.23	**-37.34**	**-37.34**	-103.77	0.18
	(106.255)	**(106.255)**	(0.479)	**(25.253)**	**(25.253)**	(4.370)	(0.394)
10	**11.58**	**11.58**	**-14.42**	195.01	195.01	-37.85	0.18
	(12.266)	**(12.266)**	**(11.683)**	(1.178)	(1.178)	(5.251)	(0.473)
11	22.58	22.58	-86.86	-1.34	-1.34	-6.97	0.18
	(5.388)	(5.388)	(1.516)	(7.540)	(7.540)	(0.797)	(0.533)

46

Table 3.9, Response (Top Row) and Relative Error (Bottom Row in Parenthesis) Using Case 4 Backcalculated Layer Model.

Point	ε_x (μm/m)	ε_y (μm/m)	ε_z (μm/m)	σ_x (kPa)	σ_y (kPa)	σ_z (kPa)	u_z (mm)
1	-58.46 (1.195)	-58.46 (1.195)	-63.09 (0.951)	-678.32 (0.055)	-678.32 (0.055)	-690.00 (0.000)	0.22 (0.860)
2	-7.08 (1.125)	-27.42 (1.238)	18.58 (1.215)	-67.82 (0.116)	-116.20 (0.155)	0.00 (0.000)	0.17 (0.790)
3	-8.48 (0.480)	-18.42 (1.114)	14.48 (0.913)	-58.02 (0.316)	-83.13 (0.046)	0.00 (0.000)	0.16 (0.730)
4	-3.30 (0.460)	-14.22 (0.985)	9.43 (0.709)	-32.18 (0.662)	-59.75 (0.196)	0.00 (0.000)	0.15 (0.671)
5	**0.00** **(100.104)**	-11.03 (0.881)	5.94 (0.495)	-15.01 (1.273)	-42.89 (0.324)	0.00 (0.000)	0.13 (0.614)
6	1.79 (3.007)	-8.63 (0.788)	3.68 (0.221)	-4.77 (3.326)	-31.10 (0.450)	0.00 (0.000)	0.12 (0.557)
7	2.75 (1.993)	-6.80 (0.700)	2.18 (0.161)	1.45 (9.928)	-22.70 (0.577)	0.00 (0.000)	0.11 (0.504)
8	22.84 (1.163)	22.84 (1.163)	-89.29 (1.051)	-70.86 (0.327)	-70.86 (0.327)	-354.14 (0.064)	0.20 (0.841)
9	**1.79** **(12.346)**	**1.79** **(12.346)**	-21.67 (1.604)	-48.76 (2.392)	-48.76 (2.392)	-108.05 (0.417)	0.19 (0.826)
10	10.60 (2.766)	10.60 (2.766)	-13.27 (2.757)	196.66 (0.342)	196.66 (0.342)	-39.78 (0.415)	0.18 (0.816)
11	21.88 (2.097)	21.88 (2.097)	-86.87 (1.524)	-1.45 (0.222)	-1.45 (0.222)	-6.98 (0.968)	0.18 (0.805)

Table 3.10, Response (Top Row) and Relative Error (Bottom Row in Parenthesis) Using Case 5
Backcalculated Layer Model

Point	ε_x (μm/m)	ε_y (μm/m)	ε_z (μm/m)	σ_x (kPa)	σ_y (kPa)	σ_z (kPa)	u_z (mm)
1	-55.77 (3.472)	-55.77 (3.472)	-57.50 (7.988)	-685.30 (1.085)	-685.30 (1.085)	-690.00 (0.000)	0.21 (1.367)
2	-7.23 (3.321)	-26.58 (1.857)	18.21 (0.793)	-68.92 (6.448)	-121.33 (4.570)	0.00 (0.000)	0.17 (0.553)
3	-7.99 (5.284)	-17.83 (2.091)	13.91 (3.102)	-59.33 (1.936)	-85.99 (3.394)	0.00 (0.000)	0.16 (0.588)
4	-3.05 (8.057)	-13.70 (2.679)	9.02 (3.705)	-32.70 (0.946)	-61.55 (2.807)	0.00 (0.000)	0.14 (0.677)
5	**0.07** **(256.180)**	-10.61 (2.996)	5.68 (3.964)	-15.20 (0.007)	-44.12 (2.545)	0.00 (0.000)	0.13 (0.755)
6	1.76 (1.183)	-8.29 (3.202)	3.51 (4.321)	-4.75 (3.761)	-31.97 (2.331)	0.00 (0.000)	0.12 (0.832)
7	2.67 (1.168)	-6.53 (3.363)	2.08 (4.825)	**1.60** **(21.260)**	-23.32 (2.138)	0.00 (0.000)	0.11 (0.906)
8	21.22 (5.981)	21.22 (5.981)	-82.82 (6.270)	-70.10 (1.401)	-70.10 (1.401)	-351.94 (0.685)	0.19 (0.820)
9	**3.35** **(109.897)**	**3.35** **(109.897)**	-21.27 (0.299)	**-36.94** **(26.065)**	**-36.94** **(26.065)**	-103.63 (4.495)	0.18 (0.554)
10	**11.63** **(12.772)**	**11.63** **(12.772)**	**-14.49** **(12.191)**	194.73 (1.320)	194.73 (1.320)	-37.81 (5.352)	0.18 (0.636)
11	22.64 (5.667)	22.64 (5.667)	-86.94 (1.614)	-1.34 (7.575)	-1.34 (7.575)	-7.00 (1.180)	0.18 (0.700)

In Tables 3.5 through 3.10, values without parentheses are the pavement responses using the backcalculated moduli in Table 3.2 (Cases 1 to 5) while values in parentheses are the relative errors compared to responses using the original (exact) pavement profile. Cells that are highlighted with gray show the response with a relative error higher than 2% whilst values in bold show the response with a relative error higher than 10%.

The results in Tables 3.3, and 3.5 through 3.9 show that even if the RMSE (of displacements) value is kept less than 1%, the resulting response of strains and stresses can largely differ than the exact response. Cases 1 and 2 (Tables 3.6 and 3.7, respectively) have relatively the same RMSE in deflection but have different relative errors in strains and stresses. In addition, the magnitude and location of the relative errors vary randomly between Cases 1 and 2. Case 1 showed a relative error of 19.686% in the horizontal strain (ε_x, or ε_y) at the bottom of the AC layer (point 9) while Case 2 showed a higher relative error of 81.365% in the horizontal strain at the same point. On the other hand, at the ground surface (point 5), Case 2 showed a relative error of 300.543% in the horizontal strain which is approximately 40 times the relative error of the horizontal strain at the same point in Case 5 (7.515%). The relative error in Case 1 was higher than 2% in 9.1% of the response points and higher than 10% in 3% of the response points; in Case 2 the relative error was higher than 2% in 48.9% of the response points and higher than 10% in 13.6% of the response points, all indicating a variable variation in the error. This high variation in the strain and stress errors can be explained by high relative error of the backcalculated elastic moduli in Case 2 as compared to that in Case 1 (Table 3.3).

Similar random variation in the relative errors can be also observed at a higher RMSE for Cases 3 and 4 as shown in Tables 3.8 and 3.9 respectively where the RMSE was almost the same (less than 1%). However, the relative error in Case 3 was higher than 2% in 65.2% of the response points and higher than 10% in 13.6% of the response points while in Case 4 the relative error was higher than 2% in 19.7% of the response points and higher than 10% in 4.5% of the response points, indicating a higher variation than Case 1. Similarly, the high variation in the strain and stress errors was attributed to the high variation in the relative error of the backcalculated elastic moduli rather than the RMSE values.

Comparing Cases 3 and 5 (Tables 3.8 and 3.10) one can observe different RMSE values but similar relative errors in the backcalculated elastic moduli. Furthermore, in both cases the

49

magnitude and location of the relative errors in the strains and stresses were nearly identical. This finding shows that controlling the displacement RMSE does not necessarily reduce the relative errors associated with the pavement response (strains and stresses) at either the surface or along the pavement profile. This would require a better control of the backcalculated moduli. However, controlling the error associated with the backcalculated elastic modulus is not an easy task since the exact modulus is not known. Even with known elastic moduli from lab testing, the variation between the backcalculated and exact moduli can be high. The difference between backcalculated subgrade elastic moduli from the FWD and laboratory elastic moduli has been studied by many researchers (Daleiden et al. 1994; Von Quintus and Killingworth 1997). It was found that there is no unique relation between the backcalculated and laboratory measured resilient moduli. The ASSHTO Guide (1993) suggested that the backcalculated modulus is three times the laboratory modulus, whilst Von Quintus and Killingworth (1997) suggested that one could use some correction factors calculated from any multilayered elastic program to match the backcalculated and laboratory moduli. However, the suggested factors are highly dependent on the backcalculation program and should be used with caution. On the other hand, Stolle (2002) showed that the moduli of the base and the subgrade layers have the largest contribution to the measured FWD deflection.

It should be noted that in all cases the effect of the backcalculated moduli on the vertical stress (σ_z) was relatively small as compared to their effect on the horizontal strains, vertical strains, and horizontal stresses.

3.3.1 Pavement Fatigue Prediction

The damage of flexible pavements can be assessed by predicting the number of loads needed to initiate cracks (fatigue cracking). The Shell Model (Bonnaure et al. 1980) and the Asphalt Institute Model (Shook et al. 1982) are frequently used for fatigue cracking in flexible pavements.

The Shell Model is based on two different loading modes, as given by Eqs. 3.3 and 3.4, below:

Shell Constant Strain Model:

$$N_\varepsilon = 13909 A_f K \left(\frac{1}{\varepsilon_t}\right)^5 E_s^{-1.8}$$

Eq. (3.3)

and Shell Constant Stress Model:

$$N_\sigma = A_f K \left(\frac{1}{\varepsilon_t}\right)^5 E_s^{-1.4}$$

Eq. (3.4)

where N_ε and N_σ are the number of load repetitions to fatigue cracking using the constant strain and constant stress analysis, respectively, A_f and K are material constants, ε_t is the tensile strain at the critical location and E_s is the stiffness of the material (i.e. elastic modulus). The constant strain model is applicable to thin AC layers usually less than 51 mm, whilst the constant stress model is applicable to thick AC layers usually more than 203 mm. The Shell Model was calibrated and generalized for any thickness as given below (MEPDG, 2004):

$$N_f = A_f K F'' \left(\frac{1}{\varepsilon_t}\right)^5 E_s^{-1.4}$$

Eq. (3.5)

where N_f is the number of load repetitions to fatigue cracking, and F'' is a constant that depends on the layer thickness and the material stiffness.

The Asphalt Institute Model is given below:

$$N_f = 0.00432 C \left(\frac{1}{\varepsilon_t}\right)^{3.291} \left(\frac{1}{E_s}\right)^{0.854}$$

Eq. (3.6)

where, similarly, N_f is the number of load repetitions to fatigue cracking, C is a material constant, ε_t is the tensile strain at the critical location, and again E_s is the material stiffness. The Asphalt Institute Model can be used for any thickness.

It can be seen from the above equations that the critical tensile strain and the stiffness of the AC layer are the main factors affecting the number of load repetitions needed to initiate fatigue failure. The effect of the backcalculated set of elastic moduli on the fatigue in flexible pavements can be studied by finding the ratio between the estimated number of repeated loads (N_f) using the backcalculated set of elastic moduli and that using the exact set of elastic moduli. In other words, the ratio is equal to N_{fb} (backcalculated set of elastic moduli) over N_{fe} (exact set of elastic moduli).

51

3.3.2 Rutting Damage

Rutting in flexible pavement is considered as a functional deterioration. Rutting is mainly predicted by calculating the vertical strains at the top of the subgrade and then estimating the allowable load repetitions until a certain rutting threshold is met. For example, Shook et *al.* (1982) assumed a rutting depth of 10 mm in their method, while Potter and Donald (1985) assumed 20-30 mm rutting depth.

Recently, the results from the test sections at MnROAD were used to develop a method to predict the number of allowable load repetitions until rutting failure using a rutting depth of 13 mm as shown in the following relation (Skok et *al.*, 2003):

$$N_r = (5.5).10^{15}\left(\frac{1}{\varepsilon_v}\right)^{3.929} \qquad \text{Eq. (3.7)}$$

where N_r is the number of allowable load repetitions until rutting failure, and ε_v is the maximum compressive strain at the top of the subgrade layer.

It can be seen, from the above equation, that the vertical strain at the top of the subgrade layer is very important to predict the lifetime of the pavement due to rutting. Similar to the fatigue case, the effect of the backcalculated elastic moduli on the rutting can be studied by finding the ratio between the estimated number of repeated loads (N_r) using the backcalculated set of elastic moduli and that using the exact set of elastic moduli. In other words, the ratio is equal to N_{rb} (backcalculated set of elastic moduli) over N_{re} (exact set of elastic moduli).

3.3.3 Fatigue and Rutting Prediction

The fatigue and rutting of the pavement are studied for the five cases as summarized in Tables 3.1 and 3.2, and the results are listed in Table 3.11. It is observed from Table 3.11 that, for all cases except Case 2, using either the Shell Model or the Asphalt Institute Model, the fatigue life will be largely underestimated based on the backcalculated moduli as compared to those based on the exact moduli. Furthermore, comparing Case 2 to Case 1 it can be seen that even for the same displacement RMSE the predicted number of repeated loads for fatigue can be largely overestimated rather than underestimated. The fatigue prediction results in Table 3.11 also show that even for a small relative error (less than 2.1% in Case 1) in the backcalculated set of elastic moduli the fatigue life of the pavement can be underestimated by 60% using the Shell model and by 45% using the Asphalt Institute Model, indicating a very high sensitivity of the fatigue life to the relative error in elastic moduli rather than the RMSE.

Table 3.11, Comparison of Fatigue Life Using Backcalculated and Exact Pavement Moduli.

Case	N_{fb}/N_{fe} Shell	N_{fb}/N_{fe} Asphalt Institute	RMSE (%)
Case 1	0.407	0.554	0.24
Case 2	4449.469	251.951	0.22
Case 3	0.027	0.092	0.65
Case 4	0.559	0.682	0.69
Case 5	0.025	0.087	0.85

The effect of the RMSE value and the relative error in pavement moduli on rutting can be observed from Table 3.12. It can be seen that even for a very small RMSE (Cases 1 and 2 where the RMSE is 0.22% and 0.24%, respectively), the rutting failure prediction based on the backcalculated moduli was overestimated by 11.7% in Case 2 while it was overestimated by 0.3% in Case 1, indicating a high sensitivity of the rutting life on the relative error in moduli. In addition, the results show that as the RMSE (error) increases, the underestimation of the rutting increases, as can be observed by comparing the rutting results in Table 3.12 for Cases 2 to 5.

Table 3.12, Comparison of Rutting Failure Using Backcalculated and Exact Pavement Moduli.

Case	N_{rb}/N_{re}	RMSE (%)
Case 1	1.003	0.24
Case 2	1.117	0.22
Case 3	0.943	0.65
Case 4	0.942	0.69
Case 5	0.939	0.85

3.4 Conclusions

This study shows that the use of the RMSE is not enough to secure an accurate backcalculation of the pavement elastic moduli. Large discrepancies can exist in the predicted pavement strains and stresses using the backcalculated and exact elastic moduli. As a result, even RMSE values less than 1% can significantly affect the fatigue and rutting predictions in flexible pavements.

The effect of the RMSE is suitable for controlling the fitness of the backcalculated deflection basin to that measured in the field while the use of the relative error in the elastic moduli is more appropriate. However, the availability of laboratory measured elastic moduli does not guarantee more appropriate backcalculated elastic moduli. This variation adds more uncertainty when dealing with data from the FWD test, and should be the future endeavor in pavement engineering.

CHAPTER IV

INTRODUCTION TO GENETIC ALGORITHMS

4.1 Introduction

The theory that proposes that organism's characteristics change over time (generations) is called the evolution theory. According to the theory, new species can be created from the parent species due to changes in the inheritable characteristics transferred by genes and chromosomes.

The theory promotes the idea of natural selection as proposed by Darwin's book (1859) *"The Origin of Species"*. In his book Darwin proposed that organisms with better characteristics survive and reproduce over generations. This means that certain characteristics can become dominant over time as a result of changes in the surrounding environmental conditions (Haldane, 1953; Lande and Arnold, 1983; Futuyma, 2005).

Mendel and later George Darwin's half-cousin Francis Galton contributed to the evolution theory by studying the mechanism of heredity in plant through experimentation and statistical means. Mendel original work was aimed at studying plant hybridization using peas. Galton, on the other hand, extended the work of Darwin to humans by creating a large database of human characteristics (including height, fingerprints, weight...etc.) to study the heredity of human traits from one generation to another. Both works in addition to the work by Darwin are considered the founding stones of the evolution theory and genetics.

Regardless of the ongoing debate concerning the theory itself, some concepts of the theory are interesting since they provide some tools and ideas to other sciences. Nature has always inspired scientists to be creative and to explore more innovative ideas and solutions to many scientific challenges. The evolution theory is no exception. Evolution can happen over billion of years while computer simulations that resemble the evolution theory can happen over few seconds to several days taking advantage of the ongoing improvements in computers.

4.2 Biological Background

All cellular forms of life are not created from nothing. These cellular forms are created based on certain rules that guarantee the same general characteristics of the base cells (parents). However, the new organisms (children) will always inherit either *the same* or *modified copies* of the parent original characteristics. Therefore, the presence of organisms with identical characteristics (even in clones) is impossible in nature. This variation is necessary in nature since species with the same characteristics transferring from one generation to the next will not survive diseases and environmental changes when they happen.

The transfer of the parent's information and the creation of new children with different characteristics are the backbones of the genetic algorithms and programming. The following sections explain the biological terms frequently used in genetic programming.

4.2.1 DNA

All genetic instructions that are needed for the formation and development of new organisms are stored in a nucleic acid known as the Deoxyribonucleic acid (DNA). The DNA is made of sugar and phosphate groups which form a skeleton that carries molecules, known as bases, connected to each other in certain sequences that are unique in each organism. The unique sequences represent the encoded genetic information that transfer from the parents to the children.

The DNA can be described as a polymer chain that consists of small repeating units which form long strands of bases that can be up to hundred of millions of base pairs. There are four main bases (nucleotides), mainly: (1) the Adenine (A), (2) the Thymine (T), (3) the Cytosine(C), (4) and the Guanine (G). These bases connect the DNA strands together. The bond between two strands consists of two different bases. On the other hand, these bases can be found in groups called codons which consist of three nucleotides and form the genetic code of the organism. Since there are 4 main bases and each codon is made of three bases, the possible number of codons is 64 ($4^3=64$).

4.3 Evolution Theory and Computer Science

Scientists were fascinated by the evolution theory not only because of its ability to explain some biological aspects of nature but because it can be used as a powerful tool in mathematics. The pioneer work of computer scientists in the 1960's and 1970's was geared toward innovative ideas for mathematical optimization similar to the work pioneered by Darwin, Mendel, and Galton. The main ideas involved creating a large number of candidate solutions (population) then applying different operators to simulate natural selection and hence to arrive at the most optimal solution for the problem in hand. This pioneer strategy was described by many computer scientists such as Rechenberg (1965) and Fogel et *al.* (1966). In modern sciences, the field of evolutionary computation is a branch of artificial intelligence. Evolution in computer science is mainly carried out using operators that simulate reproduction, mutation, recombination (crossover) and natural selection.

In addition to the use of the evolution theory as an optimization tool, scientists (Smith, 1980; Cramer, 1985; Koza, 1992) used the evolution theory as a machine-learning tool where several computer programs can be optimized using genetic programming. In genetic programming, programs can be written in any language and then natural selection and other evolution operators are applied to select the fittest program.

4.4 Genetic Algorithms

The work on evolution theory in the 1950's and 1960's was geared toward solving specific problems rather than creating generalized solution procedures. In the late 1960's John Holland was fascinated by the idea of evolution and the possibility of transforming the knowledge into computer applications. Holland (1975) was able to establish the basic theoretical foundation of the genetic algorithms by the use of the *schema* theory (plural *schemas* or *schemata*). In his original work, Holland assumed that a specie population can be represented by chromosomes which consist of strings of ones and zeros. The chromosome contains several genes (strings) that are instances of *alleles* (0 or 1). Evolution theory was then applied using natural selection and operators such as crossover, mutation, and inversion.

The distinction between Holland's genetic algorithms and the previous work in evolution theory was remarkable since Holland was able to simulate the natural selection and the evolution derived operators such as crossover, mutation, and inversion on the parent chromosomes and the offspring chromosomes as well. The previous work by Rechenberg (1965) and Fogel et *al.* (1966) used mutation to create new offspring. In addition, the population size was assumed to consist of two to few parents only.

Evolutionary algorithms (including genetic algorithms) are superior compared to the random walk methods, when dealing with complex search spaces, even though they are based on

57

random numbers. A specified domain is searched using evolutionary algorithms where "good" search points are captured and used as guidance to search for better points within the search space. However, evolution algorithms cannot be used for all problems since the availability of customized algorithms (dealing with specific problems) can be superior. The evolutionary algorithms are best suited to deal with problems where a closed form solution method is unknown or when methods cannot be used to search for an optimal solution due to the large space of the search domain.

Details of the fundamental theory of genetic algorithm and the most common operators in genetic algorithms are presented in the following sections.

4.4.1 GA Background

Genetic algorithms are robust and randomized search algorithms based on the evolution theory (natural selection) and natural genetics (Goldberg, 1989). Many calculus-based solutions are available for many problems ranging between simple to complex. However, calculus-based search methods can be inefficient when used to search for the optimal solution in a complex space since a large number of trials should be investigated in order to arrive at the most optimal solution, which makes the use of the selected method unrealistic. Other methods have been proposed for optimization using different search techniques.

The key points that make the GAs more superior than many other methods for optimizing functions of large search space and complex behavior are summarized below (Goldberg, 1989):

1-Parameters in the GAs are encoded using binary/real encoding methods and therefore the parameters themselves are not used. If the problem has more than one parameter, each parameter is encoded as a string of *l*-bit strings then a series of strings attached to each other is formed to create a *chromosome*. The entire collection of chromosomes of an organism forms the *genome*. Information encoded in the genome is called *genotype*.

2-A population of points is used to search for the optimal solution and not a single point.

3-No need for function derivatives, function continuity, changes to the optimized function, or any other auxiliary knowledge of the optimized function to perform the search. GAs use objective functions to measure the suitability of the solution (fitness). The GAs work with strings and their associated objective function values.

58

4-GAs use probabilistic transition rules, not deterministic rules.

Most of the global search methods fail to find the optimal solution especially in multimodal (many-peaked) search spaces. This can be attributed to the fact that in GA numbers are encoded as strings and therefore a population of strings is searched simultaneously rather than point by point (Goldberg, 1989). This simultaneous search technique can be described as climbing more than one peak at the same time looking for the optimal solution and hence the power of searching and the probability of finding the optimal solution are high. Each population of strings is used to create a new population of strings using genetic algorithms and the fitness of the population is evaluated by the objective function, then another population is created and so on.

The main operators that are used in a simple GA are the crossover and mutation operators. It should be noted that, even though genetic algorithms are used to *aid* in optimizing problems, some argue that they do not optimize (De Jong, 1993) but rather search the landscape problem for the best solution.

4.4.2 GA Fundamental Theory (The Schema Theorem)

The Schema Theory (plural schemata or schemas) was first introduced as the fundamental theory of genetic algorithms by Holland (1975). The Schema Theorem (or the Fundamental Theorem of Genetic Algorithms) was introduced to describe in a mathematical sense why genetic algorithms work. The *Schema Theorem* divides the search space into subspaces then quantifies the subspaces and explains the movement mechanism of individuals between subspaces. It should be noted that the standard schema theorem assumes bit flip mutation, single-cross crossover, and proportional selection.

Genetic algorithms depend mainly on a bit string representation of solutions and on *schema*. The schema is a *similarity template* that shows the similarity between different strings in the same population at certain string positions (Holland, 1975; Goldberg, 1989). Since each individual in the population is encoded using binary coding (base 2), similarities can occur between different strings that either were created randomly for the population or those that resulted from applying the natural selection operators (crossover and mutation). For example a string described by 11000 ($1\times2^4+1\times2^3+0\times2^2+0\times2^1+0\times2^0=18$) can be similar to other strings such as {11001,11011,11111,111000...etc} at the first two positions (1 and 1). Therefore, one can assume that the similar strings can follow the format of 11*** where the * indicates a "do not care" symbol. This symbol is not coded in programming the GA but it is used to illustrate the possibility of having two different values at the string position where * is encountered, namely {0,1}. Using this notation, one can create strings (schemata) using {0,1,*}. In general for a string of length (l), there are 2^l similarity templates (schemata) because each position of the template can take either its actual value or a do not care value (*). In addition, a population of size n contains between 2^l and $n.2^l$ schemata depending upon the population diversity (Goldberg, 1989). The purpose of locating similarities between strings is to guide the search besides information from the objective function (fitness function). A graphical presentation of a schema is shown in Figure 4.1. As can be seen since there are two "*do not care*" symbols (*), the number of possible new strings is $2^2=4$.

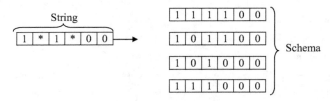

Figure 4.1, Graphical presentation of a schema example.

Not all schemata are same. Differences between schemata can be explained based on two properties; namely: (1) the order of a schema, and (2) the length of a schema. The order of a schema is equal to the number of fixed positions in the string (the number of 0 and 1), while the length of a schema is equal to the total number of positions minus the number of the "*do not care*" positions (*). The number of strings in a schema can be calculated using the following equation (White, 1993):

$$ns = 2^{l-os}$$ Eq. (4.1)

where *ns* is the number of strings in a schema, *l* is the total length of the string, and *os* is the number of the fixed positions (the number of 0 and 1)-order of schema. The defining length of a schema is the longest distance between two fixed positions.

Equation 4.1 indicates that the order of schema controls the dimension of the strings in a schema. In addition, the order of schema (*os*) plays important role in calculating the probability of survival of the schema for mutation. On the other hand, the length of a schema plays an important role in calculating the probability of survival of the schema for crossover.

Therefore, schemata are used to analyze the combined effect of the selection, crossover, and mutation on building blocks in the population. Genetic algorithms operators can disrupt the schemata depending on the length of the schema and the size of the population. In GAs, schemata of strings with high fitness values will be selected more (in a fitness-based selection) and therefore the disruption of the schemata can occur due to the mutation operator while destroying the schemata is possible due to the crossover operator. In general, the number of schemata processed in each generation with the effect of natural selection operators can be n^3.

The effect of the GAs operators on the schemata can be studied using the behavior of a schema *H* in population $x_1, x_2, \ldots _n(t)$ where *t* is a time step at which there exist $m(H,t)$ examples of a particular schema (Goldberg, 1989). In addition, assume that each chromosome C_i will be selected (fitness-based selection) with a probability $\dfrac{f(x_i)}{\sum_j f(x_j)}$. Therefore, for a population of size *n* there exists:

$$m(H,t+1) = m(H,t).n.\frac{f(H)}{\sum_j f(x_j)}$$
Eq. (4.2)

where *f(H)* is the average fitness of all strings in $x_1, x_2, \ldots _n(t)$ matching *H*. Assuming the population average to be $\bar{f} = \sum_j f(x_j)/n$, Eq. (4.2) can be written as:

$$m(H,t+1) = m(H,t)\frac{f(H)}{\bar{f}}$$
Eq. (4.3)

Assume that for a schema *H*, there will be an above-average fitness by $c\bar{f}$ amount where *c* is a constant. Therefore, Eq. (4.3) can be written as:

61

$$m(H,t+1) = m(H,t).\frac{\bar{f}+c\bar{f}}{\bar{f}} = (1+c)m(H,t) = (1+c)^{t}m(H,0) \qquad \text{Eq. (4.4)}$$

It can be seen from Eq. (4.4), even though mutation and crossover not included, that the number of trials to above-average schemata increases exponentially due to reproduction.

Crossover can either disrupt or destroy the schema. Schema disruption can occur when the crossover of the first and second mates produces a chromosome (string) with the "*do not care*" (*) at positions that can alter slightly the string rather changing the string entirely while a large change in the string value can destroy the chromosome. Schema disruption and destroying are shown in Figure 4.2 when mating strings A and H_1, and A and H_2, respectively. It can be seen, a schema can survive when the crossover position is located outside the defining length (l).

$$A = 1\,0\,0\,|\,0\,1\,1$$
$$H_1 = *\,*\,*\,|\,1\,0\,*$$
$$H_2 = *\,0\,*\,|\,*\,*\,*$$

Figure 4.2, Schema changes due to crossover.

Based on the above discussion, the survival probability under a simple crossover (p_s) can be given by (Goldberg, 1989):

$$p_s = 1 - \frac{\delta(H)}{l-1} \qquad \text{Eq. (4.5)}$$

and the probability that a schema will be destroyed (p_d) can be given by:

$$p_d = \frac{\delta(H)}{l-1} \qquad \text{Eq. (4.6)}$$

where $\delta(H) = os$ (order of schema), and l-1 is the possible positions of the cut point (crossover point) within a schema.

If crossover is performed by random choice using a crossover probability (p_c) Eq. (4.5) becomes:

$$p_s \geq 1 - p_c.\frac{\delta(H)}{l-1} \qquad \text{Eq. (4.7)}$$

Using Eqs. (4.3) and (4.7), the combined effect of the crossover and mutation, assuming independence of the crossover and mutation operations, can be expressed by:

$$m(H,t+1) \geq m(H,t)\frac{f(H)}{\bar{f}}\left[1 - p_c \cdot \frac{\delta(H)}{l-1}\right]$$
Eq. (4.8)

The above equation shows that the rate of growth or decay of schema (H) depends on a multiplication factor. Moreover, the equation indicates that schemata with above-average performance (fitness) and short defining length (l) will be sampled at exponentially increasing rates.

On the other hand, mutation can disrupt the schema since positions will be altered randomly. The mutation can affect only the fixed positions of a schema. Therefore, the disruption survival probability of a schema with a defining length $o(H)$ and probability of mutation p_m is:

$$p_t = (1 - p_m)^{o(H)}$$
Eq. (4.9)

It can be noticed that for very small values of p_m ($p_m \ll 1$), Eq. (4.9) can be approximated using:

$$p_t \approx 1 - p_m \cdot o(H)$$
Eq. (4.10)

Therefore, under the effect of reproduction, crossover, and mutation, the expected number of copies that a schema can receive in the next generation is given by:

$$m(H,t+1) \geq m(H,t)\frac{f(H)}{\bar{f}}\left[1 - p_c \cdot \frac{\delta(H)}{l-1} - o(H)p_m\right]$$
Eq. (4.11)

The above expression can lead to the Schema Theorem or the Fundamental Theorem of Genetic Algorithms. The theorem states that "Short, low-order, above-average schemata receive exponentially increasing trials in subsequent generations" (Goldberg, 1989). The theorem indicates that short, low-order schemata analyze the search space in genetic algorithms. On the other hand, Holland (1975) showed that, based on the Schema Theorem, standard genetic algorithms minimize the allocation of trials to the wrong schemata, which in turn minimize the expected loss of computation time.

The GAs schemata can be viewed as partitions of the genome space or as groups of search points that have some syntactic features.

4.4.3 Schemata Evaluation

Another advantage of GAs is the fact that the number of the evaluated schemata is larger than the number of individuals in the population (called *implicit parallelism*), which makes the GAs very efficient.

The probability that a fixed schema of order $\delta(H) = i$ matches at least one string in the population is given by:

$P(H$ matches at least one string$) = 1 - P(H$ matches none of the n strings$)$

$$= 1 - P(H \text{ matches not one string})^n$$

$$= 1 - (1 - P(H \text{ matches one string}))^n$$

$$= 1 - \left(1 - \left(\frac{1}{k}\right)^i\right)^n \qquad\qquad \text{Eq. (4.12)}$$

and the expected number of schemata of order i in n strings is:

$$\binom{i}{i} k^i \left(1 - \left(1 - \left(\frac{1}{k}\right)^i\right)^n\right) \qquad\qquad \text{Eq. (4.13)}$$

Therefore, the total number of expected schemata is:

$$S(k,l,n) = \sum_{i=0}^{l} \binom{i}{i} k^i \left(1 - \left(1 - \left(\frac{1}{k}\right)^i\right)^n\right) \qquad\qquad \text{Eq. (4.14)}$$

for an alphabet of size k, where l is the length of a string, and n is the string population size (Goldberg, 1989). When $n \to \infty$, Eq. (4.14) converges to $(k+1)^l$, which is the total number of schemata of length l over an alphabet of size k.

The use of binary encoding in GAs facilitate the use of many more schemata compared to using higher level representation which in turn covers a wider search space (Holland, 1975). The strings of a space of (n) points can be encoded as strings of length $(l = \log_k n)$ using a k-ary alphabet. As mentioned before, each string has 2^l schemata matches resulting in a total expected number of schemata equal to $(k+1)^l$ when $n \to \infty$. Both the l and the $(k+1)^l$ are maximum when $k=2$ (binary representation). The previous conclusion agrees with Goldberg's (Goldberg, 1989) principle of minimal alphabets where he suggests that the smallest alphabet should be used to permit a natural expression of the problem.

64

4.4.4 Components of Genetic Algorithms

Genetic algorithms are efficient and highly effective search and optimization tools. Genetic algorithms differ from other search and optimization algorithms and tools in the way they handle the solutions. In GAs tentative solutions (chromosomes) are encoded before being manipulated by genetic operators. Then, chromosomes are manipulated using genetic operators such as crossover and mutation. In GA function optimization, an *objective function* should be determined in order to evaluate the goodness of the individuals (chromosomes) of the population. This can be done using the *fitness function*. The fitness function in optimization problems normally is taken as the function itself.

A standard genetic algorithm works as the following (Mitchell, 1999):

1-Generate a random population of size n of l-bit chromosomes. The chromosomes represent the trial solutions to the problem.

2-Use the fitness function [$f(x)$] to calculate the fitness of each chromosome (x) in the population (n).

3-Create n offspring individuals by repeating the following steps:

i-Select two parents (chromosomes) from the current population based on the fitness values. The two parents will be biased in favor of fitness. The parent chromosomes may be selected more than once.

ii-Crossover the selected two parents to form two offspring. The crossover takes place at a randomly chosen point (using the crossover probability p_c). When no crossover is desired (crossover probability is equal to zero), the two offspring will be the exact copies of their respective parents.

iii-Mutate the two resulting offspring chromosomes at each locus with probability p_m then place the resulting chromosomes in the new population.

4-Use the new population to replace the current population.

5-Go to step 2. Repeat if necessary.

Each run of the genetic algorithm consists of many iterations where each iteration resembles a generation of chromosomes. The total number of evaluations (number of generations multiplied by the population size) is problem dependent. Genetic algorithms are randomized methods and therefore each run can result in different results and hence results are reported as average fitness of population and as average solution value(s) for a certain population size.

4.4.4.1 Population Size

A population in genetic algorithm contains a certain number of chromosomes (candidate solutions). As the number of chromosomes increases, the diversity of the population increases which in turn increases the survival of the population. The survival is attributed to the fact that a large population will have individuals with diverse characteristics that enable them to evolve and survive against any environmental changes since no dominant gene will transform from one generation to another.

However, large populations can slow down the search for the fittest (with highest possible fitness value) and can be computationally intensive. On the other hand, small populations tend to have dominant genes (genes with strong characteristics) causing early unfavorable convergence of the solution. This can be attributed to the fact that dominant solutions will restrict the search space, which will lead to local optima. Therefore, solutions in small populations will depend mainly on the mutation to produce new chromosomes that do not become dominant quickly, which in turn slows the search for the global optima.

4.4.4.2 Selection Methods

The selection of the chromosomes that will undergo crossover to produce new chromosomes (candidate solutions) can be carried out using many methods among which the "*roulette wheel*" and the "*tournament*" selection methods are the most common. In his original work Holland (Holland, 1975) used the fitness-proportionate selection where the fitness of the individual was divided by the average fitness of the population creating a "relative fitness" for each individual.

Roulette Wheel Selection

The roulette wheel selection is used to implement the fitness-proportionate selection. Each individual of the population is assigned a slice of the wheel that is proportional to the individual's fitness. Then, the wheel is spun *n* times (*n* is the population size) and the slice (and its associated individual) under the wheel marker is selected to be a parent for the next generation. However, this method can result in unlikely distribution of offspring such that all offspring are allocated to one individual due to the stochastic nature of the method.

Assume that the fitness of individual i is f_i, then the sum of all fitness values in the population (n = population size) is given by:

$$F = \sum_{i=1}^{n} f_i$$

Eq. (4.15)

Assume a running sum T_j:

$$T_j = \sum_{i=1}^{j} f_i \qquad j=1,\ldots,n$$

Eq. (4.16)

where $T_{j+1} \geq T_j \ \forall j$, and $T_n=F$.

Then generate a random number $R \in [0.0,F]$, and locate the element T_j for which:

$$T_{j-1}<R<T_j$$

Eq. (4.17)

The above procedure requires that the objective fitness value of the function to be positive definite value. The "roulette wheel" sector in this method is given a size of

$$s = \frac{1}{2\pi} \frac{f_i}{F}$$

Eq. (4.18)

while spinning the wheel is carried out using the random number generator.

In the above procedure, the probability of stopping the wheel in sector j is proportional to the angular extent of the sector, and hence the normalized fitness of individual j.

Tournament Selection

The tournament selection method (Goldberg and Deb, 1991) is more common and efficient than the roulette wheel selection method. The tournament selection method is computationally more effective since there is no need to compute the individual fitness of the chromosome and the total fitness of the population in order to implement the method. In this method, two individuals are chosen randomly from the population. A random number between 0 and 1 is selected (c) and compared to a constant parameter ($j>0.5$ in order to provide selection pressure). If the value of the random number is less than the value of the constant parameter

67

($c<j$), the individual with the higher fitness value is selected to be a parent. At the end of the process, both parents are returned back to the population with the possibility of being selected again in another iteration.

Goldberg and Deb (1991) showed that the binary tournament selection converges in $O(\log n)$ generations (n is the population size), while unscaled fitness-proportionate schemata converge in $O(n \log n)$ time.

4.4.5 Genetic Operators and Parameters

The main genetic operators and parameters are discussed in the following sections.

4.4.5.1 Elitism

In standard GAs, there is a possibility that new generations may replace good individuals with lower fit characteristics (smaller fitness) increasing the time needed to search for good individuals. This problem can be solved by using the *elitist* strategy where the best r% of the population always survives to the next generation. This is done in practice by saving in temporary storage the fittest individual of the parent population, and artificially reinserting it at the end of the generational iteration.

4.4.5.2 Crossover

The crossover operator is intended to disrupt or destroy the schemata based on the crossover strategy. As can be expected, more than one crossover strategy exists since the crossover process can be carried out by mating the chromosomes at random positions.

The simplest crossover strategy is the single-point crossover, which is based on one-point crossover operator that cuts the chromosome into two parts at random string positions. However, the single-point crossover has limited ability since in some cases it can change entirely the beneficial effect of the schemata.

The second crossover strategy is the two-point crossover, which is based on two cut points. The resulting three string parts are then combined with three other parts from another chromosome to form a new chromosome as shown in Figure 4.3. As can be seen, the single-point crossover cannot be used to combine the two schemata in Figure 4.3. However, the two-point crossover fails to work in schemata where a multiple crossover is needed to combine the chromosomes, as shown in Figure 4.4.

$$1{*}{*}{*}0 \atop {*}01{*}{*} \Big\} \quad {1|001|0 \atop 0|010|1} \Big\} \quad {1|010|0 \atop 0|001|1}$$

Figure 4.3, Example of a two-point crossover.

$$1{*}0{*}0 \atop {*}0{*}0{*} \Big\} \quad {10010 \atop 00101} \Big\} \quad {10000 \atop 00111}$$

Figure 4.4, Example where the two-point crossover fails to work.

An alternative to the single-point and the two-point crossover is the uniform crossover (shown in Figure 4.4). In the uniform crossover the location of the encoded feature on a chromosome is irrelevant compared to the single-point and two-point crossovers where the disruption possibility of the schema increases as the number of the cut positions increases. In addition, the problem of local minima in the fitness function can be avoided using the uniform crossover since good features that are encoded some distance apart on a chromosome are preserved (Syswerda, 1989).

Crossover is normally implemented using a crossover probability or rate. The crossover probability in a single-crossover represents the probability that two parents will crossover in a single point (regardless of the crossover strategy). The crossover probability in a multiple-crossover represents the number of points at which a crossover takes place.

In practice, a probabilistic test is carried out to compare between the crossover rate ($0.0 \leq p_c \leq 1.0$) and a random number generated in each crossover process ($0.0 \leq r \leq 1.0$). If the generated random number is less than the crossover rate ($r \leq p_c$), the crossover process is carried out; otherwise ($r > p_c$) the two offspring become exact copies of the two parents. For example, a value of $p_c = x$ % means that x % of the population members will participate in crossovers to

69

create new members while the remaining percent (1-x %) will be directly accepted as new members.

4.4.5.3 Mutation

Mutation of chromosomes can be carried out by flipping the bit at a random position. For example if the position contains 0, the mutation will result in a 1 at that position. In early evolution strategies, random mutation was the only source of variation while in genetic algorithms it plays a secondary (minor) role compared to the role of the crossover operator. However, in evolutionary strategy methods, mutation alone (without crossover) was used to solve complex behaviors (Ray, 1991).

In practice, a probabilistic test is carried out to compare between the mutation rate ($0.0 \leq p_m \leq 1.0$) and a random number generated in each crossover process ($0.0 \leq r \leq 1.0$). If the generated random number is less than the mutation rate ($r \leq p_m$), the mutation is carried out; otherwise ($r > p_m$) the mutation is not carried out.

4.4.5.4 Niching

Niching in biology describes the way a population responds to the surrounding resources and enemies. When there is a wealth of resources available for the population it is natural for the population to grow. On the other hand, when enemies surround the population or resources are not adequately available, the population tends to decrease. Therefore, niching explains the organism's role or job in its habitat or environment.

In genetic algorithms, niching solves the problem of focusing the search to one subdomain in the search space which in turn leads to early unfavorable convergence of the search. Early convergence in genetic algorithms can be caused by genetic drift due to stochastic errors caused by small population sizes. Therefore, it is important to give focus to other subdomains in the search space, to avoid the genetic drift, which can be done using sharing.

Modern niching schemes in genetic algorithms are based mainly on either De Jong's (1975) crowding scheme or Goldberg and Richardson's (1987) sharing scheme. In the crowding scheme, similar individuals (bit-by-bit comparison) are replaced to allow new members to enter the population. Sharing scheme uses a "sharing function" to determine the neighbouhood and

70

degree of sharing for each string in the population. The derated fitness of each individual before the selection takes place as suggested by Goldberg and Richardon (1987) is given by:

$$f_s(x_i) = \frac{f(x_i)}{\sum_{j=1}^{n} s(d(x_i, x_j))}$$
Eq. (4.19)

where $f(x_i)$ is the fitness function, and $s(d(x_i, x_j))$ is the sharing as shown in Figure 4.5.

The sharing is equal to the relative bit difference between the two chromosomes; i.e. the number of different bits (the Hamming distance between the strings) divided by the string length. Goldberg and Richardson suggested different power law sharing functions as shown in Figure 4.6. The most common power law sharing function is $(1-\Delta)$.

Figures 4.5 and 4.6 indicates that strings close to an individual (a Hamming distance of zero) require a high degree of sharing (close to one) while those far from the individual require a very small degree of sharing (close to zero). Therefore, this scheme (Eq. 4.19) limits the uncontrolled growth of particular species (a class of organism with common characteristics) within a population since individuals that are in the same neighborhood contribute to one another's share count, thus derating one another's fitness value. The niching effect is shown in Figure 4.7 for a function with equal peaks.

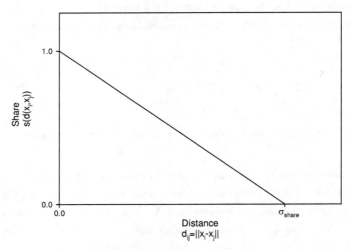

Figure 4.5, Triangular-sharing function. After Goldberg and Richardson (1987).

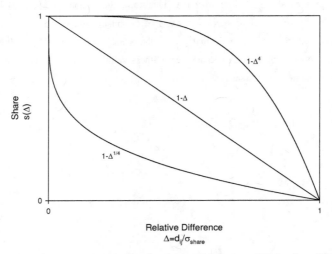

Figure 4.6, Power-sharing functions. After Goldberg and Richardson (1987).

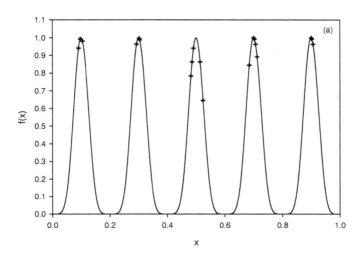

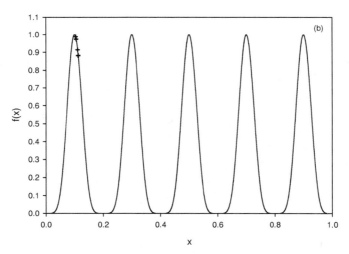

Figure 4.7, Simple genetic algorithm performance with (a) and without (b) sharing. After Goldberg and Richardson (1987).

4.5 GA Difficulties

The optimization of a function to determine the global optimum can be difficult when the function has several local optima and when the function is non-differentiable. Such functions are best optimized using genetic algorithms since no differentiability is required and functions with more than one optimum can be optimized. However, the efficiency of optimizing a function using genetic algorithms is dependent on the features of the fitness landscape and the GA parameters and operators (Holland, 1975).

Goldberg (1993) studied the difficulties that can reduce the efficiency of genetic algorithms. These difficulties include: isolation of desired solution, deception, multimodality, problem size, and the search bias.

The isolation of the desired solution can occur when the optimal solution is globally isolated and hence the surrounding points in the search landscape do not provide any information about the right search direction, and therefore, the sampling of the search space can be very difficult (Goldberg, 1989).

The optimized function can be deceptive when the resulting low order schemata, using the GA, leads the search process towards a false optimum rather than a global optimum. Deceptive functions cause undesirable convergence to local optima.

The existence of several optima in the search landscape (modality) can increase the difficulty of the GA search process for a global optimal solution. Increasing the modality of the search landscape can increase the probability of premature convergence.

The selection of the population size is very important since small populations can have low diversity and hence the premature convergence to local optima is likely, while large populations are time consuming.

Genetic algorithms search bias is dependent on the interaction between the GA parameters and the performance of the GA operators which are very complex and problem dependent.

4.6 Conclusions

Genetic algorithms are robust algorithms that can be used to optimize the search domains of functions. Genetic algorithms mimic the natural selection in nature by creating a pool of parents with characteristics that can be inherited by their children. Each individual is a string of binary code that can mate with another chromosome and mutate with time to create individuals with characteristics better than the parents.

The optimized unknown in the genetic algorithms problems can have a range rather than a seed value making the search for an optimal solution more powerful since a global solution can be obtained and divergence or local optima are prohibited when compared to other methods. In addition, genetic algorithms do not require transforming the optimized function from one form to another (such as derivative) to obtain the solution. These features of the genetic algorithms make them appealing to be used in the backcalculation of pavement layer moduli using the Falling Weight Deflectometer (FWD) test data.

CHAPTER V

BACKCALCULATION OF LAYER MODULI USING THE *BACKGENETIC3D* PROGRAM

5.1 Introduction

BackGenetic3D is a Fortran program developed by the author to backcalculate the layers' moduli of any pavement system with no restrictions regarding the number of layers, thickness of layers, location of the response points, number of loading circles, the shape of the loaded area (circle, triangle ,..etc.), and the configuration of the applied loading (uniform, nonuniform). The code consists of approximately 4100 source code lines. The program is based on the *MultiSmart3D* forward multilayer elastic program developed by the Computer Modeling and Simulation Group at the University of Akron (Pan and Alkasawneh, 2006). The *MultiSmart3D* program is based on the innovative computational and mathematical techniques for multilayered elastic systems (e.g., Pan, 1989a,b, 1990, 1997).

BackGenetic3D is a very advanced and fast program that is capable of backcalculating the elastic moduli accurately. The program searches the domain of the elastic moduli for possible solutions (moduli) and determines the optimal solution using a guided stochastic search technique. The search technique is based on the genetic algorithm technique proposed by Holland (1975) with many improvements to handle the complexity associated with the backcalculation of layers' moduli. The implemented search technique is designed to avoid the local optima that can be encountered when using other methods such as the Simplex method.

The program is user-friendly and can be used with the typical inputs that are used in any elastic backcalculation program. In the *BackGenetic3D* program there is no need for seed moduli and a range of the seed moduli is required while in other programs, that do not use the GA, a seed modulus is needed for each layer. Using seed moduli can lead to local optima or can lead to divergence rather than convergence.

The genetic operators needed to perform the backcalculation should be provided by experienced users or can be estimated using special techniques as those suggested in this book. More details regarding the capability and important guidelines on how to use the

BackGenetic3D program in backcalculating the elastic moduli are discussed throughout this thesis.

5.2 BackGenetic3D Program Components

The ***BackGenetic3D*** program consists of many components that work together to backcalculate the elastic moduli. The backcalculation procedure starts with the selection of the parameters of the unknown layers including the Poisson's ratio, layer thickness, and typical ranges of the elastic moduli based on the description of the layer (material type). Figure 5.1, shows the main components of the program and the sequence of the components. As can be seen, the program follows the natural selection by creating a pool of parents then mating the parents to create new offspring (chromosomes). The fittest

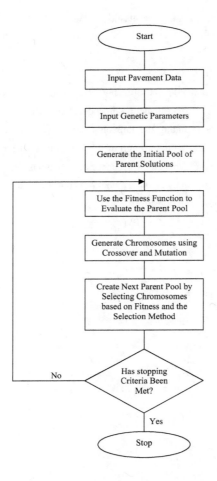

Figure 5.1, Flow chart of the genetic algorithm implemented in the ***BackGenetic3D*** program.

offspring are then selected as the best candidates that can survive the surrounding environment represented by the constraints (moduli ranges) and the fitness function. The fitness function facilitates the selection by providing a feedback about the quality of each offspring and if it will be beneficial to keep it for the next generation to form the new parent pool. The stopping criterion in *BackGenetic3D* is either the generation number or the fitness value.

5.3 Backcalculation Mechanism

The program starts by randomly generating the initial pool of parents using a random number generator, which is desirable since no bias will exist among the selected individuals and the generation will cover a wide range of the search domain. This approach is different than the available backcalculation programs that do not use the GA where seed moduli should be provided in order to start the backcalculation process. Selecting the seed moduli can create a bias toward one or all of the selected moduli, which can be far from the optimum solution. The program uses Mersenne Twister random generator (Matsumoto and Nishimura, 1998) to create the parents pool.

The individuals form the genotype. The genotype is a chromosomal presentation of the solutions where each genotype consists of genes. Each chromosome contains the trial solutions encoded in a binary string format. For example, if the pavement system contains three layers, there will be three unknowns and hence the chromosome will consist of three genes (moduli). The program deals with chromosomes when executing the genetic algorithm rather than genes. Figure 5.2, represents a chromosome that contains three genes which is used to solve a three-layer pavement system.

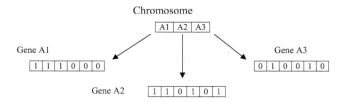

Figure 5.2, Chromosomal presentation of the trial solutions encoded as binary bits.

79

5.4 Population Size

The size of the parent pool represents the size of the generation in the genetic algorithm and is determined by the user. The size of the generation is used as an input parameter by the random generator to create the initial parent pool. The generation can be of any size depending on the users experience and on the complexity of the problem. However, increasing the population size will increase the computational time compared to smaller population sizes. It is recommended to use large population sizes as increasing the size will increase the diversity and the distribution of the parent pool. Diversity of the population is highly recommended for pavement moduli backcalculation since a wider range of the search domain will be covered and hence the probability of getting stuck with local optima will be minimized and the need for high selection pressures through the genetic operators will be reduced.

Goldberg and Deb (1991) and Goldberg et al. (1992) suggested using the following equation to determine the size of the population (n) assuming that the function can be viewed as the sum of m independent subfunctions:

$$n = O(m\chi^k)$$ Eq. (5.1)

where $m = l/k$, l is the total number of bits in a chromosome, χ is the alphabets cardinality, and k is the order of the schemata which can be approximated as the string length of each gene (modulus). Using binary coding, the above equation can be reduced to:

$$n = O(\frac{l}{k}2^k)$$ Eq. (5.2)

The effect of the number of layers on the population size assuming $k=2,4,8,10$ is shown in Figure 5.3. As can be seen, the population size increases as the number of layers increases and as the binary string length of genes increases indicating that the population size is problem dependent. However, the figure indicates the worst-case scenario and does not necessarily indicate the population size required to backcalculate the moduli. Experience with pavement moduli backcalculation, in this book, showed that the required population size is less than those shown in Figure 5.3 by many folds.

80

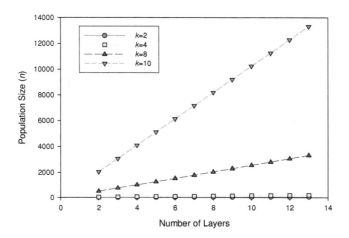

Figure 5.3, Population sizing relations as a function of number of layers and binary string length of genes.

5.5 Crossover Probability

Crossover in genetic algorithms is a very important operator since it is responsible for combining the good characteristics of the existing individuals (chromosomes) and hence works as features enhancer. Different crossover techniques can be used in genetic algorithms including the single-point crossover, the two-point crossover, and the uniform crossover. The single-point and the uniform crossovers are implemented in **BackGenetic3D**. In the single-point crossover a point is selected at random along the chromosome length, then two children (offspring) are produced by selecting and combining one part from each parent. If the single point-crossover is used, it is recommended to use high probability values (more than 0.5) to ensure more diverse population at the initial stages of the backcalculation and hence avoid premature convergence. In general, values less than 0.6 are rarely used (Eiben et *al.*, 1999).

81

The uniform crossover allows a variable number of crossover points and therefore avoids any bias toward any of the string bits. Syswerda (1989) showed that the uniform crossover is better than the single-point and the two-point crossover. In addition, he argued that the schema in the uniform crossover get destroyed at a faster rate than in the single-point and two-point crossovers and get created at a far faster rate.

5.6 Mutation

Mutation was normally used as the main operator in evolution strategies where the diversity of the population is controlled by one factor. In genetic algorithms, mutation allows the production of new chromosomes with new characteristics that are different from those in the parents. Mutation ensures a controlled diversity in the population to avoid premature convergence. On the other hand, mutation is one of the means by which genetic algorithms exploit the search domain for the best solutions.

BackGenetic3D uses jump mutation where the probability of mutating each chromosome is equal to the mutation probability. In mutation, a binary string in the chromosome is flipped to its opposite value. For example, if the mutated binary string is 1 it will be flipped to 0 while if the string is 0 it will be flipped to 1. By flipping binary strings, the value of the gene changes dramatically. When the mutation probability is equal to one (P_m=1.0), all chromosomes will be mutated resulting in a complete random (not guided) generation of chromosomes. It is recommended to use low mutation probabilities since diversity is desirable without the destruction of the schema. Mutation can be viewed as an occasional alteration of the chromosomes. It is common to use a mutation probability equal to either the inverse of the population size or the inverse of the chromosome length.

5.7 Creep Mutation

The *BackGenetic3D* program is the first genetic algorithm program to use creep mutation in the backcalculation of pavement moduli. In creep mutation the values of the chromosomes are altered slightly by shifting the values backwards or forwards to the neighboring value. The rate of change imposed on the chromosomes using the creep mutation is less than that using the uniform mutation. It should be noted that the creep mutation acts on the decoded gene (phenotype) while the mutation works on the encoded chromosome (genotype). A common value for creep mutation is equal to the ratio between the length of the gene binary string to the population size.

5.8 Selection Procedure

The selection procedure is very important in genetic algorithms since it determines the mating chromosomes for the crossover process in order to create new chromosomes (solutions). The *BackGenetic3D* program uses the tournament selection method. In tournament selection, two chromosomes are chosen at random and then compared based on their fitness values. Using other selection methods such as the roulette wheel selection, the linear ranking selection, and steady state selection were excluded. Goldberg and Deb (1991) showed that the tournament selection is highly efficient and requires less population sizes to converge compared to many other selection methods suggested in literature. In the tournament selection a significant selection pressure is applied while the pitfalls of fitness ordering or ranking are avoided (Falkenauer, 1998).

5.9 Niching

Niching is used for the first time in the backcalculation of pavement moduli in *BackGenetic3D*. Niching is a user defined option and can be either used or turned off. Niching helps in preserving temporally useful diversity. Niching is helpful when the size of the population used is tends to cause a premature convergence. In such situations, niching is highly recommended

The implemented niching in the *BackGenetic3D* program is based on Goldberg and Richardson's (1987) sharing scheme which is more appropriate for multimodal functions. Sharing scheme uses a "sharing function" to determine the neighbouhood and the degree of sharing of each string in the population. The derated fitness of each individual before selection takes place as suggested by Goldberg and Richardon (1987) is given by:

$$f_s(x_i) = \frac{f(x_i)}{\sum_{j=1}^{n} s(d(x_i, x_j))} \qquad \text{Eq. (5.3)}$$

83

where $f(x_i)$ is the fitness function, and $s(d(x_i, x_j))$ is the sharing. The sharing is equal to the relative bit difference between the two chromosomes; i.e. the number of different bits (the Hamming distance between the strings) divided by the string length.

Goldberg and Richardson suggested different power law sharing functions. The most common power law sharing function is $(1-\Delta^\beta)$. The value of β is user defined in **BackGenetic3D**. However, a value of β=1.0 is commonly used and is the default in **BackGenetic3D**.

5.10 Elitism

Elitism strategy is used for the first time in the backcalculation of pavement moduli in **BackGenetic3D**. Elitism is a user defined option and can be either used or turned off. In the elitism strategy, the chromosome with the best fitness is always retained. Therefore, elitism can only be used after calculating the fitness of the chromosomes. Elitism is beneficial in reducing the required number of generations since convergence to the best solution is achieved quickly. In addition, it is recommended to use elitism especially when the selected genetic operators tend to disrupt the schema of the new chromosomes such as when the mutation probability is high.

5.11 Population Seeding

The initial population (parent pool) is generated randomly at the start of each run. However, **BackGenetic3D** has the capability of seeding the population using information from a previous run. In this technique, the last chromosomes are retained externally and used as the initial parent pool for the next generation. Population seeding is a user defined option and can be either used or turned off. The default option is no population seeding.

5.12 Random Number Generator

The genetic algorithms are based on generating stochastic random numbers to spatially distribute the parent pool. In addition, random numbers are used during many steps in the genetic algorithm computation process. Therefore, it is important to ensure that the used random generator is highly stochastic and the probability of generating the same number or gene is extremely low.

The *BackGenetic3D* program uses Mersenne Twister (MT) random generator as proposed by Matsumoto and Nishimura (1998). The MT random generator employs a powerful and fast algorithm that is capable of generating very high quality pseudonumbers. The algorithm was designed to solve many of the disadvantages associated with other random number generators. The MT algorithm is very powerful since it has a proved and extremly large period of $(2^{19937}-1)$ and the probability of duplicating the value is negligeble. On the other hand, the algorithm is suitable to be used in computer implementations when speed is of a concern since the generation of random numbers is fast. The seed value of each random generation is automated and obtained using the clock of the computer (time, day, and year) to gurantee complete randomness during the run.

5.13 BackGenetic3D Program Verification

The validity of the *BackGenetic3D* program has been assessed by comparing its performance with the performance of other available programs and backcalculation procedures. A robust backcalculation program should include a robust backcalculation procedure to minimize or eliminate (if possible) the problems of local optima and seed moduli effect on the backcalculation results. In reality, all of the popular commercially available backcalculation programs share the similar backcalculation procedure. However, the main key in these programs is the estimation of the seed moduli in order to minimize the effect of the seed moduli on the backcalculated moduli realizing the drawback of the seed moduli and the premature convergence to local optima.

A four-layer composite pavement system was used to validate the backcalculation procedure as shown in Table 5.1. The pavement system was analyzed by Harichandran et *al*. (1993). Harichandran et *al*. (1993) backcalculated the pavement moduli using three backcalculation programs (MICHBACK, MODULUS, EVERCALC). MICHBACK was developed by Harichandran et *al*. (1993) while MODULUS and EVERLCALC were developed by Uzan et *al*. (1989) and Sivaneswaran et *al*. (1991), respectively. The backcalculation was performed using a deflection basin generated using a load of 40.03 kN applied to a circular plate of radius 150.1 mm. The deflection basin consisted of seven surface deflections at radial distances of 0,203.2, 304.8, 457.2, 609.6, 914.1, and 1524 mm from the center of the loaded plate.

Table 5.1, Four layer pavement system.

Pavement Layer	Thickness (mm)	Poisson's Ratio	Actual Modulus (MPa)
AC	152.4	0.35	3,447.38
Slab	254.0	0.25	31,026.40
Base	203.2	0.4	172.37
Subgrade	—	0.45	51.71

In addition to the above programs, **BackGenetic3D** was used to backcalculate the pavement moduli using the GA parameters shown in Table 5.2 and the pavement moduli ranges shown in Table 5.3.

Table 5.2, Range of GA inputs used in the four-layer pavement.

Parameter	Value
Population size	512
Number of generations (G)	150
Probability of crossover (P_c)	0.5
Probability of mutation (P_m)	0.0

Table 5.3, Ranges of the moduli.

Layer	Moduli (MPa)
AC	1,000-6,000
Slab	10,000-60,000
Base	100-400
Subgrade	10-150

86

The GA fitness function was assumed to be:

Minimize: $$RMSE = \sqrt{\frac{1}{m}\sum_{i=1}^{m}\left(\frac{d_i - D_i}{D_i}\right)^2}$$ Eq. (5.4)

where $RMSE$ is the root mean square error, m is the number of measuring sensors, d_i is the backcalculated deflection at point i, and D_i is the measured deflection at point i. In addition, the relative error (RE) of each backcalculated modulus was estimated using:

$$RE = \frac{E_c - E_m}{E_m} \times 100\%$$ Eq. (5.5)

where E_c and E_m are the backcalculated and the actual moduli, respectively.

The relations between the number of generations and the best fitness and the associated RMSE of the generation are shown in Figure 5.4 while the relative error of each modulus is shown in Table 5.4.

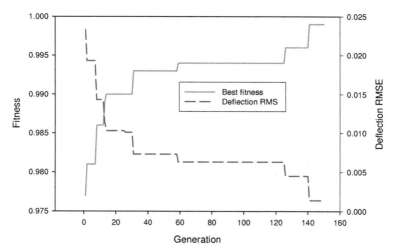

Figure 5.4, Best fitness and deflection RMS relations with the number of generations for the 4-layer pavement system.

Table 5.4, Results of the backcalculation programs.

Program	AC (MPa)	Slab (MPa)	Base (MPa)	Subgrade (MPa)	Deflection RMSE (%)
MICHBACK	3443.03 (0.13)*	31131.9 (0.34)*	158.59 (7.99)*	51.75 (0.08)*	0.007
MODULUS	3639.74 (5.58)*	30827.1 (0.64)*	67.57 (60.80)*	52.40 (1.33)*	0.068
EVERCALC	10908.93 (216.44)*	15838.0 (48.95)*	90.95 (47.24)*	51.83 (0.23)*	1.526
BackGenetic3D	3445.656310 (0.05)*	31036.018184 (0.031)*	172.800925 (0.25)*	51.606580 (0.2)*	0.0014

* Relative error (%) of the backcalculated moduli.

It can be seen from Figure 5.4 and Table 5.4 that the BackGenetic3D was able to optimize the entire search domain to locate the optimal solutions. In addition, the results show that even that the other programs were able to achieve the commonly recommended RMSE of less than 3%, the discrepencies between the actual and the backcalculated moduli can be extremly high. The backcalculation results of EVERCALC showed the highest discrepancy with an error of 216.44% in the top layer. In addition, the results indicate that even if the RMSE is less than 1% the discrepancy can be high due to the local optima effect as shown by MODULUS results where the RMSE is 0.068 while the modulus relative error of the base layer is 60.80%.

5.14 Conclusions

A new pavement backcalculation program is developed based on the genetic algorithms method. The new program implements many new operators and parameters that have been used for the first time to backcalculate the pavement moduli. In addition, a powerful and robust random number generator (Mersenne Twister Random Number Generator) was implemented as well.

The verification problem showed that the BackGenetic3D program is superior than many other programs that backcalculate the pavement moduli using classical backcalculation methods. The BackGenetic3D program is designed to optimize the entire search space looking for the global optima rather than the local optima.

CHAPTER VI

OPTIMIZATION OF GENETIC ALGORITHMS OPERATORS AND PARAMETERS FOR

THE BACKCALCULATION OF THE PAVEMENT ELASTIC MODULI

6.1 Introduction

Carrying out a successful genetic algorithm run to find the optimal parameters of a function requires the use of the genetic algorithm operators properly and the selection of the right population size. Genetic algorithm operators and strategies include crossover, mutation, niching, elitism, and many other operators. The selection of the genetic algorithm parameters has not been generalized in the form of rules or guidelines. This can be attributed to the nature of each problem where the operators and strategies work as tools that help in optimizing the search space.

The genetic algorithm operators and strategies can differ from one problem to another. The interaction between the parameters can be one of the reasons why the same operators cannot be generalized to all problems. In biology, two genes can have two different effects at different loci in the genome; however, the two genes can still be strongly linked. Interaction between genes can lead to the manifestation of certain characteristics of the organism (epistatic gene) or the suppression of another gene's characters (hypostatic gene). In genetic algorithms, the same can be observed when the optimized function contains many parameters interacting with each other (especially nonlinear functions).

On the other hand, parameters of the optimized function are encoded as strings of l-bit which combined consist one chromosome. Therefore, assuming a problem with three parameters where the string length of each parameter is 18 bits, the total length of the chromosome will be 54 bits (18 bits x 3 parameters). It is known that the size of the search space for a bit string encoding of length l is 2^l (Goldberg, 1989), which forms a hypercube (l-dimensional hypercube) where the genetic algorithm samples the corners of the hypercube (Whitley, 1994).

The search space size can be very large based on the exponential relation 2^l. For example, Winston (1992) pointed out that the effective size of the search space of possible board configurations in chess is in the order of 2^{400}. This large number is extremely impractical and hence random search techniques should be used. Therefore, the biased random search in genetic algorithms is suitable for such cases. The selection of the genetic operators and strategies will play a large role in directing the search in very large or small search spaces depending on the size of the space.

On the other hand, the mechanism of each genetic algorithm operator affects the value of the parameter itself. For example, the values of the crossover operator when single-crossover or multiple-crossover methods are used will be different for the same problem. Therefore, when operators are used, the value of the operator should be taken into account together with the method (strategy) used to implement that operator.

The interaction of different techniques with the GA operators can affect the value of the operators. For example, the implementation of elitism and niching (sharing) (Goldberg, 1989; Goldberg and Richardson, 1987) can largely affect the values of the GA operators.

The complexity of the search space is another important factor that affects the value of the operator. For example, searching the space of multimodal functions depends on the complexity of the function. Multimodal functions are functions with more than one optima with the possibility of having a single or more than one global optima with same value. Thus, in constrained search spaces, the degree of modality (number of optima) depends on the window (limits) of the search space, as shown in Figure 6.1. For example, if the search space in a region is to be investigated, the value of the crossover operator will be different than that applied to search another region within the same search space. On the other hand, single or double crossover operations can be destructive on individuals (solutions) near different optima.

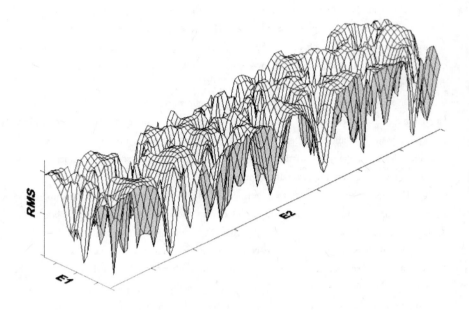

Figure 6.1, Search Space of a 2-Layer Pavement Section.

6.2 Selection of GA Operators

Designing a GA to optimize a problem requires the selection of the crossover and the mutation operators, the size of the population, and the number of generations. These requirements impose a dilemma to the users since no standard method or guidelines are available. However, several attempts have been carried out since Holland (Holland, 1975) placed the foundations of the GAs.

The first systematic work to study the interaction between the GA operators was carried out by De Jong (1975). De Jong used 5 test functions as shown below:

$$F1(X) = \sum_{i=1}^{3} x_i^2 \quad , -5.12 \le x_i \le 5.12 \qquad \text{Eq. (6.1)}$$

$$F2(X) = 100(x_1^2 - x_2)^2 + (1 - x_1)^2 \quad , -2.048 \le x_i \le 2.048 \qquad \text{Eq. (6.2)}$$

$$F3(X) = \sum_{i=1}^{5} \text{ integer}(x_i) \quad , -5.12 \le x_i \le 5.12 \qquad \text{Eq. (6.3)}$$

$$F4(X) = \sum_{i=1}^{30} ix_i^4 + Gauss(0,1) \quad , -1.28 \le x_i \le 1.28 \qquad \text{Eq. (6.4)}$$

$$F5(X) = \left[0.002 + \sum_{j=1}^{25} \frac{1}{j + \sum_{i=1}^{2}(x_i - a_{ij})^4} \right]^{-1} \quad , -65.536 \le x_i \le 65.536 \qquad \text{Eq. (6.5)}$$

In his work, De Jong (1975) used roulette wheel selection, single-point crossover, and simple mutation. It was concluded that, though, small populations can lead to quick results, large populations are more recommended since they can lead to more accurate results. The GA parameters were recommended based on the performance of all of the five test functions (Eq. 6.1 through Eq. 6.5). The recommended values for crossover probability (per pair of parents) and mutation probability (per bit) were 0.6 and 0.001, respectively, while the recommended population size ranged between 50 and 100. In addition, De Jong recommended the use of low mutation rates all the time. The recommended mutation probability was equal to the inverse of the population size ($p_m = 1/n$).

Grefenstette (1986) extended the work of De Jong (1975) by optimizing the GA parameters of the five functions (Eq. 6.1 through Eq. 6.5) using the GA as an optimization tool. Grefenstette (1986) used elitism to enhance the search ability. The recommended values for crossover probability (per pair of parents) and mutation probability (per bit) were 0.95 and 0.01, respectively, while the recommended population size was 30. Grefenstette (1986) recommended the use of higher crossover and mutation probabilities, which contradict the recommended values by De Jong (1975). Grefenstette (1986) concluded that higher crossover probabilities (close to $p_c = 1.0$) can be used when more accurate selection procedures are used to reduce the stochastic errors of sampling.

Schaffer et al. (1989), studied many test functions including some of De Jong's (De Jong, 1975) test functions to provide recommendations on the GA parameters. The recommended

93

values for crossover probability and mutation probability were 0.75-0.95 and 0.005-0.01, respectively, while the recommended population size ranged between 20 and 30. However, Schaffer et *al.* (1989) concluded that the selected GA parameters were independent of the tested functions.

Other studies included recommendations on the selection of some of the GA parameters. Mühlenbein (1992) and Bäck (1992), working separately, recommended the use of a mutation probability equal to the inverse of the string (chromosome) length ($p_m=1/l$). This recommended value is best suited for ($1+1$) GAs where each single parent chromosome produces a single child by means of mutation. Deb (2005) suggested using crossover values between 0.7 and 1.0.

As can be seen from the above review, different studies suggested different parameter sets. None of the studies was able to show that the suggested parameters are the optimal parameters that can be used for any problem. However, many GA users use these parameters as a rule of thumb for optimization. It should be emphasized that manual tuning of parameters is required especially for functions that have not been studied before.

6.3 Backcalculation of Pavement Moduli

The backcalculation of the pavement moduli involves the use of forward multilayer elastic programs. Trial moduli are used to calculate the deflections at the Falling Weight Delectometer (FWD) sensors and then the fitness function is used in the genetic algorithm to quantify the quality of the backcalculated moduli. The search space in the pavement backcalculation problem is multimodal and hence the required genetic parameters should be dealt with care. The use of genetic parameters from studies other than those dealing with the same problem can be misleading since the complexity of the problem and the size of the search space are unique compared to the studied functions in many investigations.

Optimizing the pavement backcalculation problem is challenging since the size of the search space can be very large compared to search spaces in standard genetic algorithm optimized functions such as those studied by De Jong (1975), Grefenstette (1986), and Schaffer et *al.* (1989) where the search space was limited to one region. In the pavement backcalculation problem, each modulus has its own search region making the interaction between the studied regions an extra challenge to the genetic algorithm method. Pavement backcalculation problems

94

include, normally, at least two unknown elastic parameters (elastic moduli of layers) while other parameters are known or assumed (layer thicknesses, Poisson's ratios).

On the other hand, the selection of the search region is not the same for all moduli (all elastic layers in the pavement system). The search space (region) for each modulus should take into account the physical characteristics and the type of the layer in the system. For example, the search space for the concrete-pavement layer will be different than that of the base, subbase, and the subgrade layers due to differences in the physical properties that change the elasticity characteristics of each layer. However, due to the dependency of the elastic parameters on temperature/moisture and mechanical distresses in the pavement (micro/macro cracks), the search space should be expanded to account for any possible moduli that solve the problem. This in turn increases the effort required to solve the problem since more modal points will exist in the search space. Therefore, one can expect different genetic parameters for different search spaces. Figure 6.1 shows the search space for a 2-layer pavement system over the selected search range. As can be seen from the figure, the difficulty of locating the best solution (optimal solution) is largely affected by the search window within the search space. As the search window increases the difficulty increases.

6.4 Genetic Operators Interaction and Performance

The backcalculation of the pavement layers moduli using genetic algorithms is very robust and effective if the right genetic operators are used. The selection of high or low operators can highly influence the convergence of the objective function and hence all genetic operators should be chosen carefully. For example, by definition, the selection of high mutation rates (probabilities) can increase the randomness of the backcalculated solutions (diversity of the chromosomes/individuals) due to the high changing rate of the binary code that represents the parameters. On the other hand, the selection of very low mutation rates can reduce the change in the chromosomes characteristics and hence will reduce the convergence rate to the best solution (best fit of the objective function).

However, the mutation rate is not the only operator that affects the performance and the behavior of the backcalculation process using genetic algorithms. The effect of each parameter and the interaction between the parameters cannot be studied explicitly using mathematical formulations due to the complexity of the problem and the complex search space of the backcalculated parameters. Therefore, the interaction between the genetic operators can be done more practically using sensitivity analysis.

The objectives of the scarce available pavement backcalculation researches using genetic algorithms were aimed at demonstrating the applicability of the method (Fwa et *al.*, 1997;Kameyama et *al.*, 1997; Reddy et *al.*, 2004; Tsai et *al.*, 2004). In these researches, the selection of the genetic parameters was based on trial and error (in Reddy et *al.*, 2004, 1200 trials were used) to optimize the selected operators. The selected operators were chosen based on the value of the fitness function and the previous knowledge of the pavement layer moduli. The fittest solution was determined by comparing the backcalculated moduli with the known moduli.

However, such approach suffers from several major drawbacks. The main objective of any backcalculation procedure is to find the fittest parameters (optimum layer moduli) where the parameters are unknown beforehand. The second drawback is the lack of any guidelines on how to make the selection of the genetic operators more practical rather than carrying out several economically inefficient and time consuming runs since, in practice, the only reference for solution quality control is the displacement objective function.

Another drawback is the generalization of one-problem results to other problems. For example, the results of De Jong's (1975) test functions cannot be used to optimize all other functions and problems since different search spaces will result in different complexities. In addition, moduli backcalculation using genetic algorithms showed different optimum genetic operators, as shown in Table 6.1. It can be seen that the population size range is 60-500, the generation range is 40-593, the mutation probability range is 0.02-0.15, and the crossover probability range is 0.85 to 0.9. The inconsistency in the genetic parameters is evident regardless of the number of backcalculated moduli (number of layers). The computational effort in Table 6.1 was computed, as suggested by Deb (1995), for the optimum genetic operators. The computational effort is a measure of the required effort and time by each set of genetic parameters to converge to the best solution.

96

Table 6.1, Summary of Pavement Backcalculation Studies.

Study	Population Size	No. of Generations (G)	No. of Layers	Mutation Probability (P_m)	Crossover Probability (P_c)	CE*
Fwa et al. (1997)	60	40-150	3	0.15	0.85	5.1-15.3
Fwa et al. (1997)	60	120	4	0.15	0.85	15.3
Kameyama et al. (1997)	50	42-593	4	Variable	Variable	NA
Reddy et al. (2004)	100	60	3	0.02	0.90	1.08
Reddy et al. (2004)	60	60	3,4	0.10	0.74	4.44
Tsai et al. (2004)	500	50	3	NA	NA	NA

*Computational Effort: CE=G×P_m×P_c

Reddy et al. (2004) optimized the genetic operators using a 3-layer pavement system (100,60,0.02,0.90). In their work, additional analysis using the 3-layer system was carried out to estimate the optimum genetic operators (60,60,0.10,0.74) that can be used for 3- and 4-layer systems. The resulting two different sets of genetic operators show that the selected genetic operators are not unique even for the same problem and search space and they may not be optimal. The computational effort (CE) of the two sets can be 4.11 times the other (4.44/1.08). On the other hand, for the same number of layers, results from Fwa et al. (1997) and Reddy et al. (2004) indicated that the computational effort can be 3.44 times a possible available solution (15.3/4.44).

The above conclusions and comparisons assume that the search will converge to the best solution. Fitting the displacement functions was the main objective during the genetic algorithms iterations while the optimum solution was selected by comparing the backcalculated moduli and the original known moduli of the pavement. The fitness functions used in all studies are shown below:

Fwa et al. (1997): Minimize $RMSE = \sqrt{\dfrac{1}{m}\sum_{i=1}^{m}\left(\dfrac{d_i - D_i}{D_i}\right)^2}$ Eq. (6.6)

Kameyama et al. (1997): Minimize $RMSE_k = \sum_{i=1}^{m}\sqrt{w_i(d_{ik} - D_{ik})^2}$ Eq. (6.7)

Reddy et al. (2004): Minimize $RMSE = \sum_{i=1}^{m}(d_i - D_i)^2$ Eq. (6.8)

Tsai et al. (2004): Minimize $RMSE = \sum_{i=1}^{m}(d_i - D_i)^2$ Eq. (6.9)

where $RMSE$ is the root mean square error, m is the number of measuring sensors, d_i is the backcalculated deflection at point i, D_i is the measured deflection at point i, w is the weight factor at point i, and k is the chromosome number.

As can be seen from the above equations, the fitness function in all studies was in general the same.

6.5 Sensitivity Analysis

Sensitivity analysis was carried out to study the effect of the genetic operators on the search space and to study the interaction between the genetic operators to aid in the selection of the parameters even if the modulus of the elasticity is unknown beforehand, which is the goal of the backcalculation analysis and the common case in practice.

The sensitivity analysis was carried out using the problem studied by Reddy et al. (2004), shown in Figure 6.2. Reddy et al. (2004) investigated the backcalculation of the elastic moduli of a 3-layer pavement system using different trials to obtain the optimum genetic operators. A total of 1200 parameter sets were created in their study using the ranges shown in Table 6.2. The optimum parameters shown in Table 6.2 are the parameters of the best solution. The best solution is defined as the solution with the highest fitness and the lowest average moduli error. The fitness function used by Reddy et al. (2004) was minimized using:

Maximize $f = \dfrac{1}{1 + RMSE}$ Eq. (6.10)

where f is the fitness of the solution based on the root mean square error as estimated in Eq. (6.8). The *RMSE* was calculated for six points along the surface of the pavement located at 0,0.3,0.6,0.9,1.2, and 1.5 m from the center of the loading plate.

The effect of each genetic operator was studied by varying one of the operators while the others are constant. Two populations (6 and 100) were used during the analysis to study the effect of the population size on the operators as well. The number of generations was selected to be 200 for the small population size and 100 for the large population size, resulting in a total number of evaluations equal to 1,200 and 10,000, respectively. The range of the parameters used in the sensitivity analysis is shown in Table 6.3. The length of each chromosome was equal to 45.

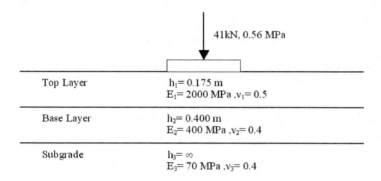

Figure 6.2, The 3-layer pavement system used in the sensitivity analysis, after Reddy et *al.* (2004).

Table 6.2, Range of GA inputs used by Reddy et *al.* (2004).

Parameter	Range	Optimum
Population size	20-160	100
Number of generations (G)	20-160	60
Probability of crossover (P_c)	0.6-0.95	0.9
Probability of mutation (P_m)	0.001-0.2	0.02

Table 6.3, Values of parameters used in the sensitivity analysis.

Parameter	Values
Population size	6,100
Number of generations (G)	200*,100**
Probability of crossover (P_c)	0.0-1.0
Probability of mutation (P_m)	0.001,0.01,0.1,0.5,0.7,1.0

* Small population.

** Large population.

As can be seen the crossover probability values ranged between no crossover (P_c= 0.0) to 100% crossover (P_c=1.0) and the mutation values probability values ranged between very slow mutation (P_m= 0.001) to 100% mutation (P_c=1.0).

The "goodness" of the selected parameters using the fitness function given by Eq. 6.10 was further assessed using the following equation:

$$RMSE_E = \sqrt{\sum_{i=1}^{m}\left(E_{ci} - E_{mi}\right)^2}$$
Eq. (6.11)

where $RMSE_E$ is the root mean square error of the backcalculated moduli of elasticity, and E_{ci} and E_{mi} are the backcalculated and the pavement system moduli (actual moduli), respectively, for layer i.

Moduli ranges of the studied pavement system used in the backcalculation procedure are shown in Table 6.4.

Table 6.4, Moduli ranges of backcalculated moduli.

Layer	Moduli Range (MPa)
Top Layer	1200-2800
Base Layer	240-560
Subgrade Layer	40-100

6.5.1 Behavior of Small Populations

The crossover probability was varied between 0.0 and 1.0 for the small population size (population size=6) as shown in Figures 6.3 through 6.13. Each figure shows the relations between the generation average fitness and the $RMSE_E$ of the backcalculated moduli under a certain set of mutation and crossover rates. The generation average fitness explains the average convergence behavior of the backcalculation procedure.

Figure 6.3a shows the effect of the mutation rate on the generation average fitness (will refer to as the fitness) when the crossover rate is equal to zero, which implies no crossover during the evaluations (runs). It can be seen that as the mutation rate increases, in this example, the fitness decreases due to the increase of the randomized flipping of the digits of the chromosomes, which in turn randomizes the backcalculated solutions (moduli) and hence it increases the diversity of the chromosomes. When the mutation rate is equal to one a complete "chaos" is produced since all chromosomes will be subjected to mutation and hence the new generation will be completely different than the old generation and none of the fittest chromosomes (solutions) will be preserved for the next generation. In addition, decreasing the mutation rate increased the fitness since variation was controlled to a minimum level and hence premature (early false) convergence to the best solution (since no crossover) was somehow minimized.

According to De Jong (1975) the optimal mutation probability is equal to the inverse of the population size, which is equal to 0.167 or 1/6. Mühlenbein (1992) and Bäck (1992) recommended an optimal mutation probability equals to the inverse of the chromosome length, which is 0.02 or 1/45. It can be observed that for the studied problem (3-layer pavement system) the random effect of the mutation rate on the fitness was less observed when the mutation rate was less than 0.1. Mutation rates less than 0.01 showed the highest fitness values compared to all other mutation rates. The best mutation rate of the studied set of parameters in Figure 6.3a was 0.001, which is 1/20 times the recommended value by Mühlenbein (1992) and Bäck (1992), and 1/167 the recommended value by De Jong (1975).

However, Figure 6.3b shows a different effect of the fitness function on the root mean square error of the backcalculated moduli ($RMSE_E$). The results suggest that the possibility of having $RMSE_E$ values close to zero (perfect match between the backcalculated and the model

101

elastic moduli) increases as the mutation rate increases which contradicts the fitness values behavior in Figure 6.3a where the fitness decreased as the mutation rate increased. In addition, a mutation rate of 0.01 showed the highest $RMSE_E$ error while a mutation rate of 1.0 showed the highest possibility of zero $RMSE_E$ error. This behavior suggests that high fitness values (Eq. 6.10) can be associated with "incorrect" backcalculated moduli. This shows that different sets of layers moduli can be far from the optimal layers moduli but still at the same time satisfy the fitness function.

On the other hand, Figures 6.4 through 6.7 show the effect of the variation of the mutation rate while the crossover rate is constant (P_c =0.25,0.5,0.75, and 1.0). In general, increasing the mutation rate beyond 0.01 increased the randomness of the fitness values while mutation probabilities less than 0.01 exhibited higher fitness values (Figures 6.4a through 6.7a). In addition, increasing the crossover rate reduced the $RMSE_E$ error regardless of the mutation rate (Figures 6.4b through 6.7b). It is evident from the figures that the mutation rate can enhance the values of the fitness and the $RMSE_E$ error when the crossover is more than 0.5.

Figure 6.8 presents the relation between a constant very low mutation rate (Pm=0.001) and variable crossover rates ranging between zero and one. A uniform crossover probability was used as suggested by Spears and De Jong (1991) in which an exchange happens at each bit position with a probability of P_c. It can be seen that increasing the crossover probability for a mutation probability of 0.001 increased the fitness values for all of the P_c values. The fittest solutions (using Eq. 6.10) had the highest $RMSE_E$ errors (Figure 6.8b), with values up to 60% of the backcalculated moduli. The lowest $RMSE_E$ errors were obtained when the mutation probability was 0.001 and the crossover probability was higher than 0.75 (0.75 and 1.0). This indicates that, in small populations, high crossover probabilities are more appropriate when the mutation probability is very small.

Increasing the mutation probability more than 0.001 resulted in more variation of the chromosome bits and hence increased the diversity of the new generations as shown in Figures 6.9a through 6.13a where the fitness variation did not follow a certain trend. High mutation rates reduced the transfer of the elite chromosomes (chromosomes with high fitness) from one generation to another. In addition, the variation of the $RMSE_E$ errors increased as the crossover probability increased while the mutation probability was constant (Figures 6.9b through 6.13b).

102

In general, the $RMSE_E$ errors decreased regardless of the crossover probability when the mutation probability was higher than 0.1. On the other hand, the fitness decreased regardless of the crossover probability when the mutation probability was higher than 0.1. These behaviors suggest again that the use of the fitness function alone in practice can be deceptive and hence convergence to the optimal solution can be difficult.

It can be seen that for small populations, there is a need to increase the mutation probability since more diversity can be induced when the mutation probability increases. However, when the mutation rate is equal to 1.0, a complete random generation of chromosomes will be produced.

When the population size is small the possibility that similar genes dominate the gene pool increases due to the lack of diversity among the genes and hence the possibility of the premature convergence increases. In general, when the crossover probability is zero and the mutation probability is very small, the GA search becomes a hill-climbing search. On the other hand, when the crossover probability is zero and the mutation probability is very high, the GA search becomes a random search.

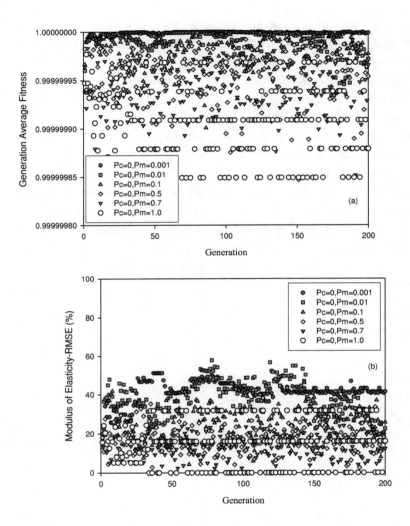

Figure 6.3, Generation average fitness (a) and RMSE of the backcalculated moduli (b) under
variable mutation rates and constant crossover rate (P_c=0.0) for a small population.

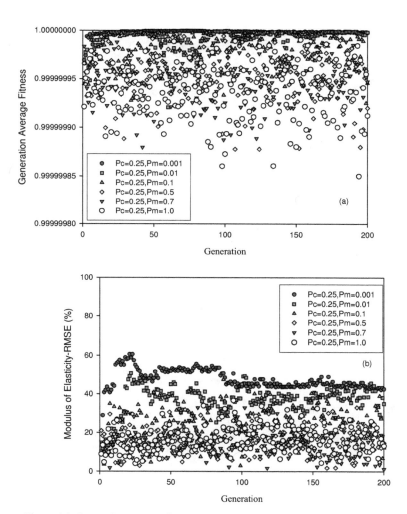

Figure 6.4, Generation average fitness (a) and RMSE of the backcalculated moduli (b) under variable mutation rates and constant crossover rates (P_c=0.25) for a small population.

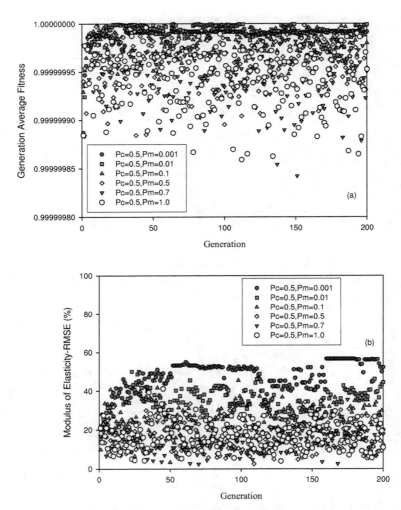

Figure 6.5, Generation average fitness (a) and RMSE of the backcalculated moduli (b) under variable mutation rates and constant crossover rates (P_c=0.5) for a small population.

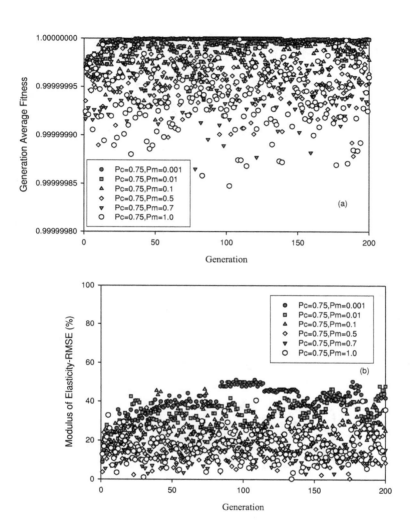

Figure 6.6, Generation average fitness (a) and RMSE of the backcalculated moduli (b) under variable mutation rates and constant crossover rates (P_c=0.75) for a small population.

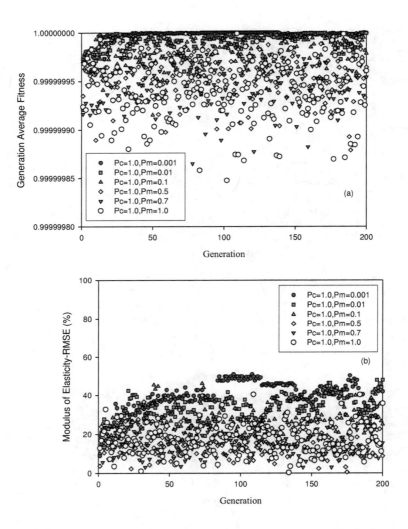

Figure 6.7, Generation average fitness (a) and RMSE of the backcalculated moduli (b) under variable mutation rates and constant crossover rates (P_c=1.0) for a small population.

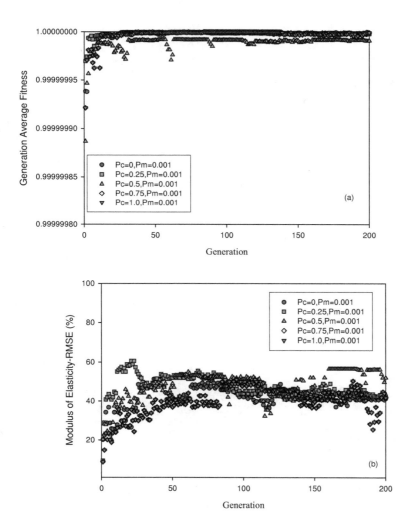

Figure 6.8, Generation average fitness (a) and RMSE of the backcalculated moduli (b) under constant mutation rates (P_m=0.001) and variable crossover rates for a small population.

109

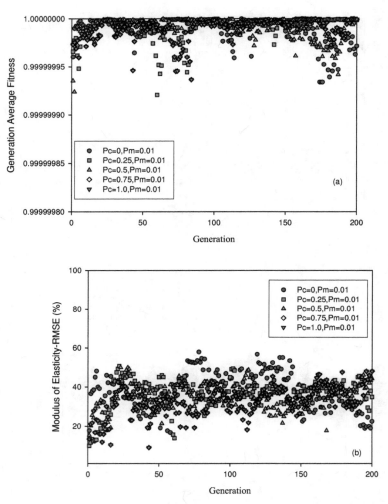

Figure 6.9, Generation average fitness (a) and RMSE of the backcalculated moduli (b) under constant mutation rates (P_m=0.01) and variable crossover rates for a small population.

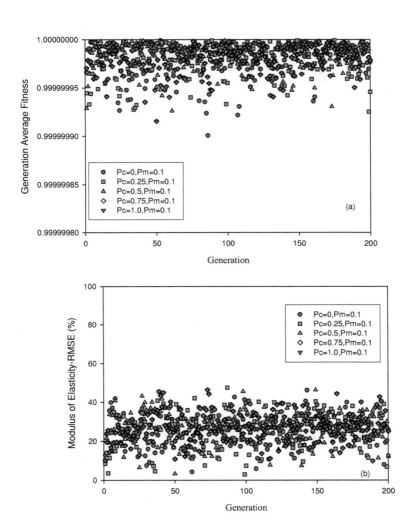

Figure 6.10, Generation average fitness (a) and RMSE of the backcalculated moduli (b) under constant mutation rates (P_m=0.01) and variable crossover rates for a small population.

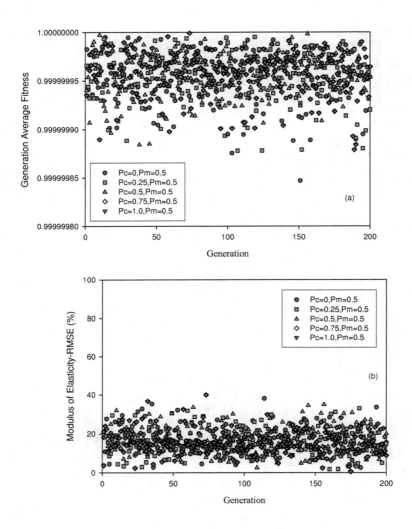

Figure 6.11, Generation average fitness (a) and RMSE of the backcalculated moduli (b) under constant mutation rates (P_m=0.5) and variable crossover rates for a small population.

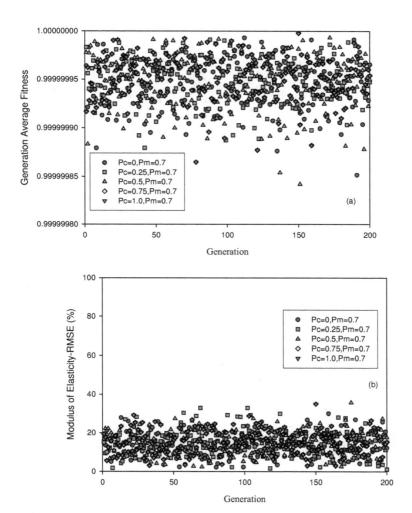

Figure 6.12, Generation average fitness (a) and RMSE of the backcalculated moduli (b) under constant mutation rates (P_m=0.7) and variable crossover rates for a small population for a small population.

113

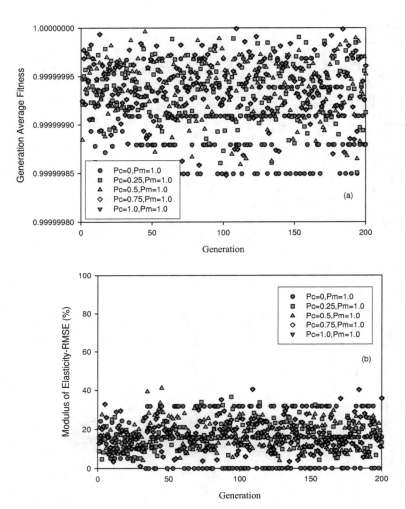

Figure 6.13, Generation average fitness (a) and RMSE of the backcalculated moduli (b) under constant mutation rates (P_m=1.0) and variable crossover rates for a small population.

114

6.5.2 Behavior of Large Populations

The effect of the population size was further studied using a large population size of 100 and a total number of generations of 100 (total evaluations=10,000). The crossover probability was varied between 0.0 and 1.0 for the large population size as shown in Figures 6.14 through 6.24. Each figure shows the relation between the generation average fitness and the $RMSE_E$ of the backcalculated moduli under a certain set of mutation and crossover rates.

It can be seen that increasing the population size (approximately 16.67 times the small population size) reduced the effect of the mutation on the fitness. In Figure 6.14a, the fitness changed slightly compared to the change in fitness in Figure 6.3a (small population). In addition, in this example, the effect of large mutation probabilities increased as the number of generations increased, indicating a destructive effect of high mutation probabilities on the schema of the chromosomes. The same destructive effect of high mutation probabilities can be seen in small populations as well (Figure 6.3a). However, the rate of schema destruction in small populations is higher than that in large populations. On the other hand, the small change in fitness (in Figures 6.14 through 6.24) in large populations lead to a relatively (about 10% change in $RMSE_E$) large change in the backcalculated moduli indicating high sensitivity to the change in the fitness. Furthermore, the $RMSE_E$ error sensitivity to the change in fitness is higher in small populations compared to that in large populations.

Figures 6.14 through 6.18 show that increasing the crossover in large populations can increase the fitness values and, at the same time, can reduce the associated $RMSE_E$ errors. However, the effect of increasing the crossover was pronounced when the crossover probability was less than 0.5 while the effect became negligible with probabilities higher than 0.5 and less than 1.0. When the crossover probability is 1.0, all individuals in the population will be subjected to crossover. The change in the GA performance due to the change in the crossover probability is attributed to the change in the selection pressure. It should be noted that, in tournament selection two individuals are selected and compared based on the fitness values and the fittest is selected for mating. In addition, when the mutation probability is equal to 1.0, a complete random generation of chromosomes will be produced.

Figures 6.15 and 6.16 shows that the best performance of the GA can be achieved (for a population size=100) when the crossover probability was between 0.25 and 0.5 while the mutation probability was less than 0.01.

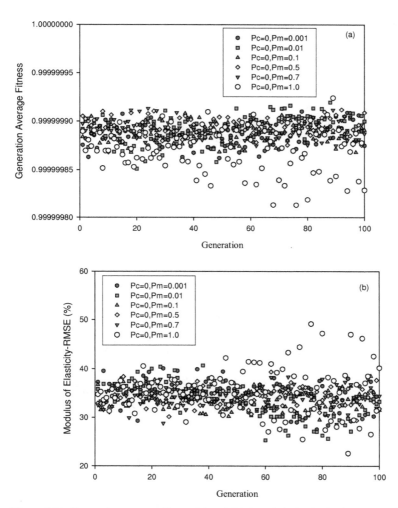

Figure 6.14, Generation average fitness (a) and RMSE of the backcalculated moduli (b) under variable mutation rates and constant crossover rates (P_c=0.0) for a large population.

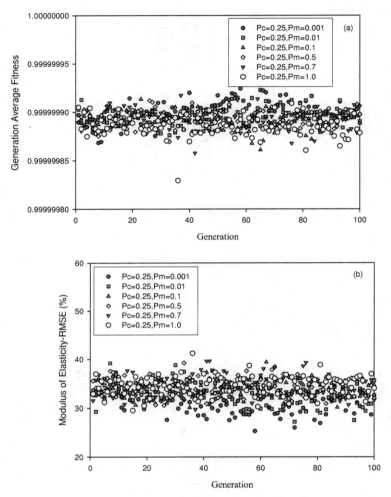

Figure 6.15, Generation average fitness (a) and RMSE of the backcalculated moduli (b) under variable mutation rates and constant crossover rates (P_c=0.25) for a large population.

118

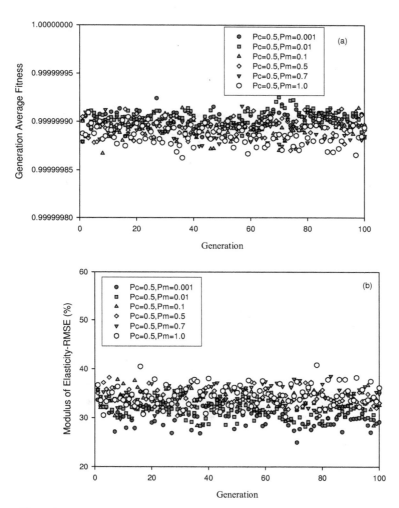

Figure 6.16, Generation average fitness (a) and RMSE of the backcalculated moduli (b) under variable mutation rates and constant crossover rates (P_c=0.5) for a large population.

119

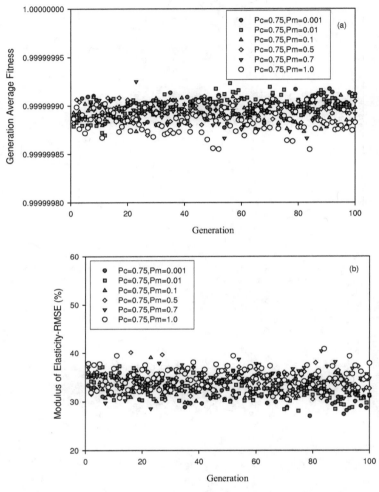

Figure 6.17, Generation average fitness (a) and RMSE of the backcalculated moduli (b) under variable mutation rates and constant crossover rates (P_c =0.75) for a large population.

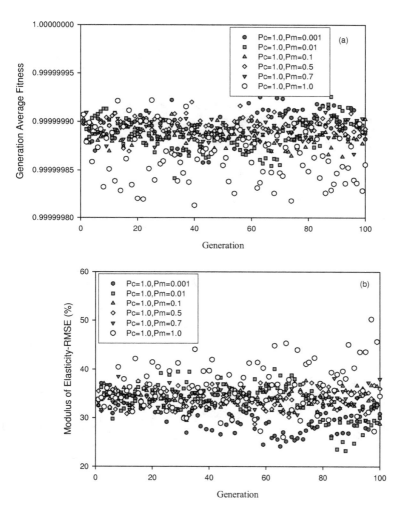

Figure 6.18, Generation average fitness (a) and RMSE of the backcalculated moduli (b) under variable mutation rates and constant crossover rates (P_c =1.0) for a large population.

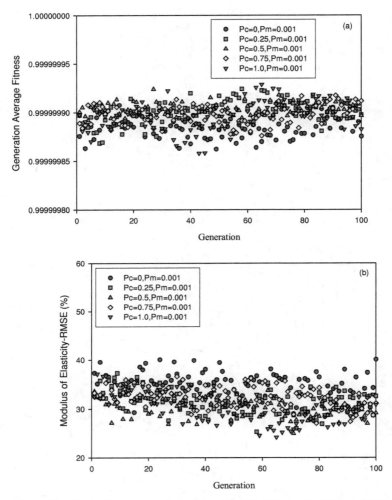

Figure 6.19, Generation average fitness (a) and RMSE of the backcalculated moduli (b) under
constant mutation rates (P_m=0.001) and variable crossover rates for a large population.

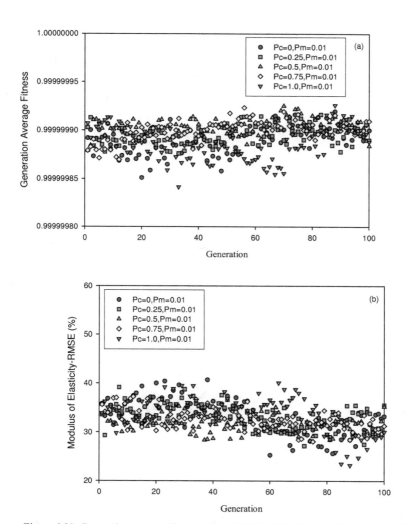

Figure 6.20, Generation average fitness (a) and RMSE of the backcalculated moduli (b) under constant mutation rates (P_m=0.01) and variable crossover rates for a large population.

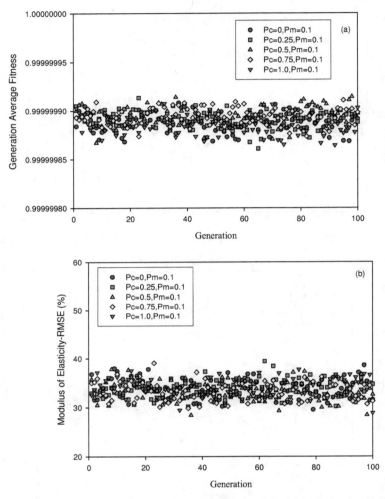

Figure 6.21, Generation average fitness (a) and RMSE of the backcalculated moduli (b) under constant mutation rates (P_m=0. 1) and variable crossover rates for a large population.

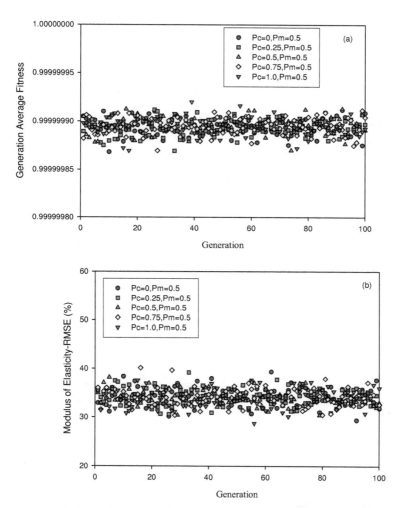

Figure 6.22, Generation average fitness (a) and RMSE of the backcalculated moduli (b) under constant mutation rates ($P_m=0.5$) and variable crossover rates for a large population.

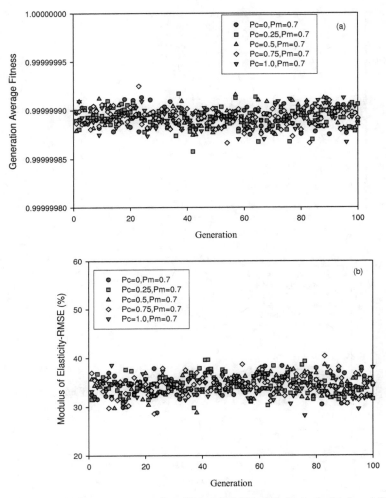

Figure 6.23, Generation average fitness (a) and RMSE of the backcalculated moduli (b) under
constant mutation rates (P_m=0.7) and variable crossover rates for a large population.

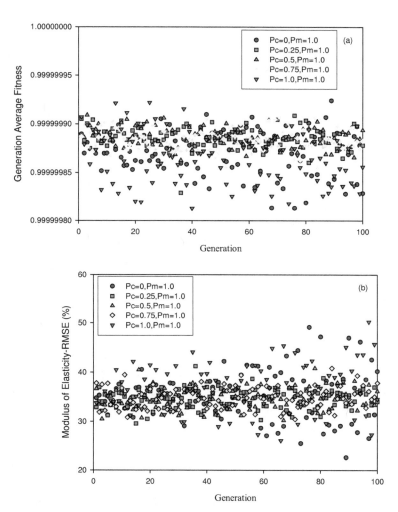

Figure 6.24, Generation average fitness (a) and RMSE of the backcalculated moduli (b) under constant mutation rates (P_m=1.0) and variable crossover rates for a large population.

6.5.3 Behavior of Large Populations Using Different Ranges of Moduli

The effect of the search domain on the performance of the backcalculation process was further studied using different ranges of pavement moduli, as shown in Table 6.5. The large population size was compared according to the change in moduli ranges to further understand the complexity of the search domain. Figures 6.25 through 6.35 show the results of the new ranges of moduli.

Table 6.5, Moduli ranges of backcalculated moduli.

Layer	New Moduli Range (MPa)	Old Moduli Range (MPa)
Top Layer	800-3200	1200-2800
Base Layer	160-640	240-560
Subgrade Layer	28-112	40-100

It can be seen that changing the moduli ranges by increasing the search domain of each moduli increased the fitness and reduced the associated $RMSE_E$ error. In general the $RMSE_E$ errors reduced by approximately 85% by changing the ranges of the moduli while the increase in the fitness values was relatively small.

Figures 6.14 and 6.25 indicate different performances of the GA. In Figure 6.14b at low mutation probabilities (less 0.1), and while there is no crossover (P_c=0), the effect of the mutation was almost negligible on the fitness values and moderate on the associated $RMSE_E$ error while in Figure 6.25b the effect was moderate and high, respectively.

Figures 6.26 through 6.29 indicate that the fitness values increase and the associated $RMSE_E$ errors decrease by increasing the crossover probability. On the other hand, Figures 6.26 through 6.29 and Figures 6.15 through 6.18 show different responses to the mutation and crossover probabilities. In Figures 6.26 through 6.29 increasing the mutation increased the fitness and reduced he associated $RMSE_E$ error while in Figures 6.15 through 6.18, the increase in the mutation probability had the opposite effect by reducing the fitness and increasing the associated $RMSE_E$ errors. The same observations can be drawn by comparing Figures 6.30 through 6.35 and Figures 6.19 through 6.24.

The above example shows that due to the complexity of the search domain and the interaction between the GA parameters and operators, recommending one set than can be used for any problem is not feasible. In addition, the selection of the moduli range is as important as selecting the genetic operators and parameters.

Increasing the range of the unknown pavement moduli is recommended since the algorithm will be able to search for the optimal solution within a wide range of solution (as in the current ranges of moduli) rather than getting stuck with a subdomain (as in the previous ranges of moduli) leading to incorrect conclusions and backcalculated moduli. However, increasing the moduli ranges will increase the effort of the GA and can lead to different results than desired. The physical characteristics of the layers should be used as guidelines when selecting the appropriate ranges of moduli. Furthermore, Table 6.5 can be used as a guideline for the selection of the appropriate ranges of moduli based on the description of the pavement material. However, based on the author's experience and to give more flexibility to the GA search, Table 6.6 included some recommended ranges to be used in the backcalculation of the pavement moduli using GA.

Table 6.6, Typical values of modulus of elasticity for pavement materials.

Material	Range* (MPa)	Recommended Range (MPa)
Hot-Mix Asphalt	1,500-3,500	1,000-4,000
Portland Cement Concrete	20,000-55,000	18,000-60,000
Asphalt-Treated Base	500-3,000	300-3,5000
Cement-Treated Base	3,500-7,000	2,500-8,000
Lean Concrete	7,000-20,000	6,000-25,000
Granular Base	100-350	80-450
Granular Subgrade Soil	50-150	30-250
Fine-Grained Subgrade Soil	20-50	10-100

* After (AASHTO, 1993).

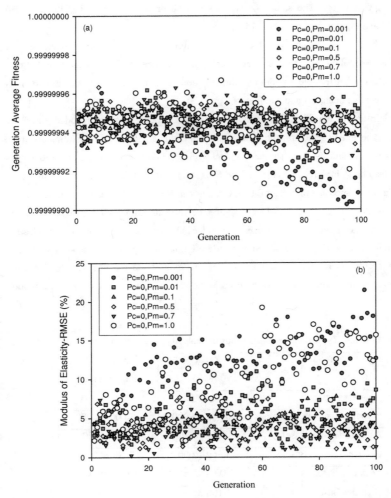

Figure 6.25, Generation average fitness (a) and RMSE of the backcalculated moduli (b) under variable mutation rates and constant crossover rates (P_c=0.0) for a large population (Different ranges of moduli).

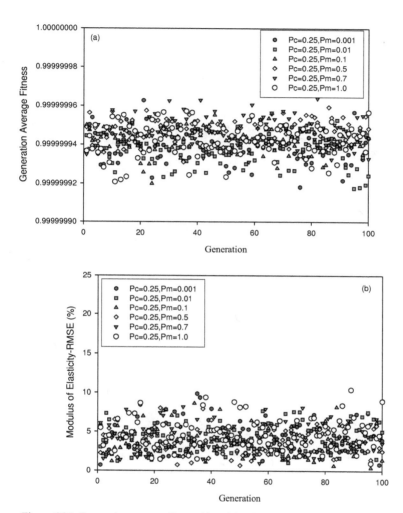

Figure 6.26, Generation average fitness (a) and RMSE of the backcalculated moduli (b) under variable mutation rates and constant crossover rates (P_c=0.25) for a large population. (Different ranges of moduli).

131

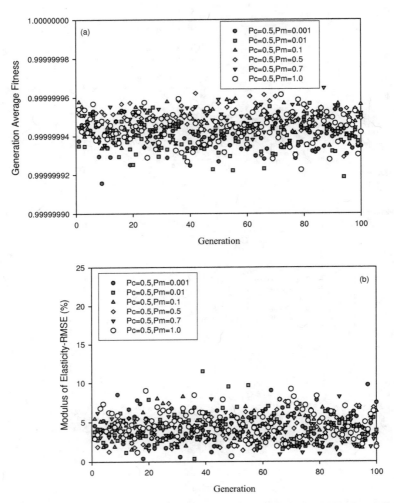

Figure 6.27, Generation average fitness (a) and RMSE of the backcalculated moduli (b) under variable mutation rates and constant crossover rates (P_c=0.5) for a large population. (Different ranges of moduli).

132

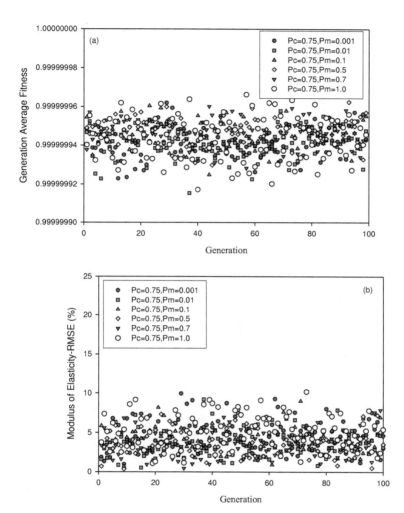

Figure 6.28, Generation average fitness (a) and RMSE of the backcalculated moduli (b) under variable mutation rates and constant crossover rates (P_c =0.75) for a large population. (Different ranges of moduli).

133

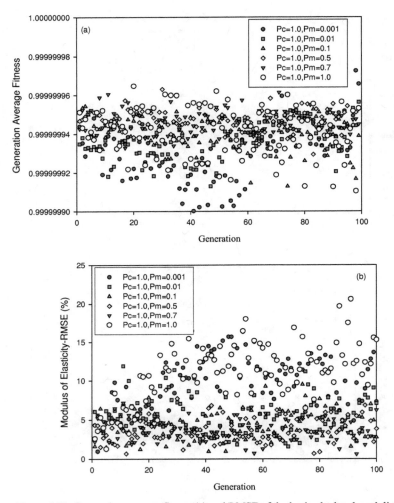

Figure 6.29, Generation average fitness (a) and RMSE of the backcalculated moduli (b) under variable mutation rates and constant crossover rates (P_c =1.0) for a large population. (Different ranges of moduli).

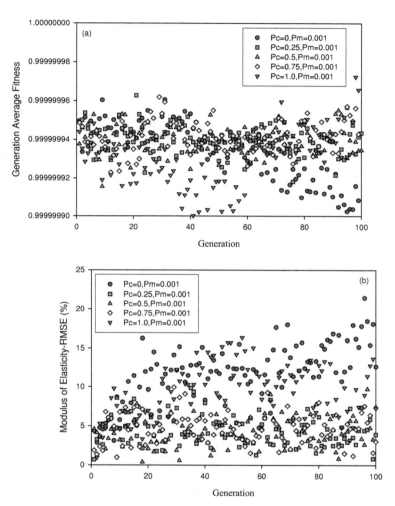

Figure 6.30, Generation average fitness (a) and RMSE of the backcalculated moduli (b) under constant mutation rates ($P_m=0.001$) and variable crossover rates for a large population. (Different ranges of moduli).

135

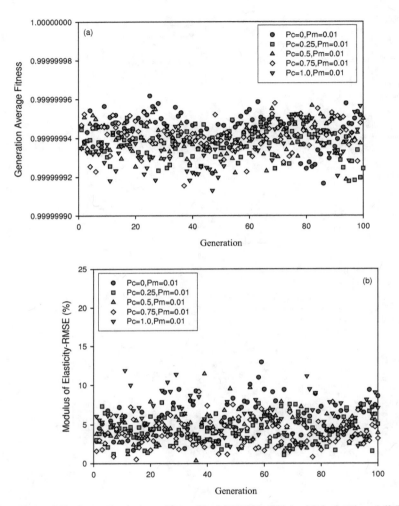

Figure 6.31, Generation average fitness (a) and RMSE of the backcalculated moduli (b) under constant mutation rates (P_m=0.01) and variable crossover rates for a large population. (Different ranges of moduli).

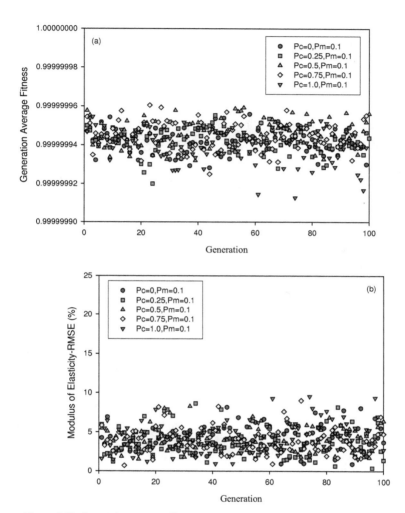

Figure 6.32, Generation average fitness (a) and RMSE of the backcalculated moduli (b) under constant mutation rates (P_m=0. 1) and variable crossover rates for a large population. (Different ranges of moduli).

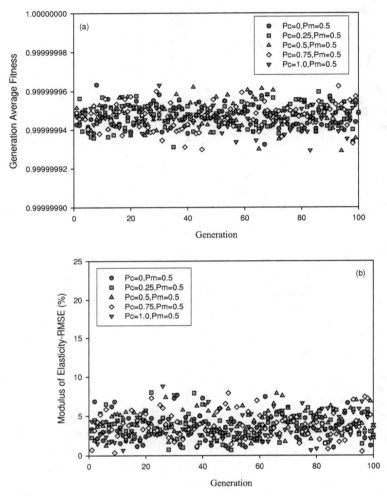

Figure 6.33, Generation average fitness (a) and RMSE of the backcalculated moduli (b) under constant mutation rates (P_m=0.5) and variable crossover rates for a large population. (Different ranges of moduli).

138

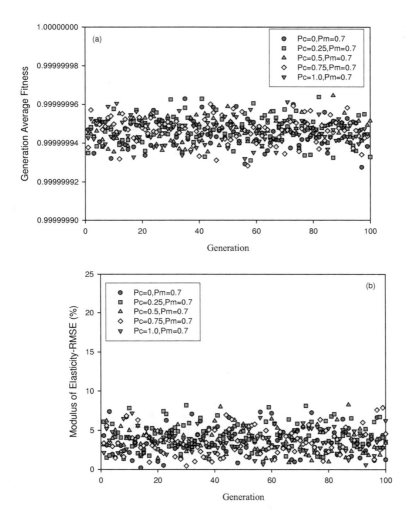

Figure 6.34, Generation average fitness (a) and RMSE of the backcalculated moduli (b) under constant mutation rates (P_m=0.7) and variable crossover rates for a large population. (Different ranges of moduli).

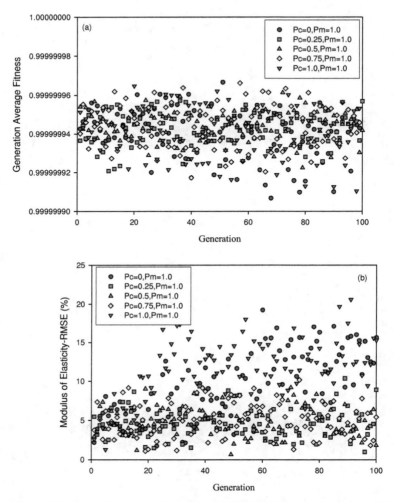

Figure 6.35, Generation average fitness (a) and RMSE of the backcalculated moduli (b) under constant mutation rates (P_m=1.0) and variable crossover rates for a large population. (Different ranges of moduli).

6.5.4 Best Fitness and Average Fitness Comparison

The use of the best fitness instead of the average fitness in the sensitivity analysis to optimize the selection of the genetic parameters can be argued to be less favorable. Understanding the meaning of both the average fitness and the best fitness can help in understanding the advantages and the disadvantages of using both in the sensitivity analysis. The best fitness value represents the fitness of the fittest chromosome (moduli) within the population while the average fitness value represents the average of all fitness values (high and low) within the population. The selection process in the genetic algorithms favors the chromosomes with high fitness values, therefore, the fittest chromosome will be transferred (sometimes with the help of elitism) from one generation to the other. However, the crossover and the mutation processes will work together to produce chromosomes with higher fitness values from the selected chromosomes. This indicates that the fitness of the individuals within the population will increase with time and one of the best solutions will dominate the population after several generations an can lead to local convergence if the population size is small. The convergence to a solution does not mean always, as was shown previously, that the fittest chromosome is the "correct" solution even if the fitness value is high.

Therefore, if one uses the best fitness as the only measure to select the genetic algorithms parameters, the best fitness can be misleading since it will show the performance of the best chromosome only while the mechanism of the search process and the interaction between the genetic parameters can not be studied. In addition, since the genetic algorithms method works as a guided stochastic search technique, it is more appropriate to study the behavior of averages rather than single values. The average fitness value shows the performance of the entire population and how the selection, crossover, and mutation processes interact with each other within the current search space. When the best fitness used in the sensitivity analysis, the selection of the number of generations and the population size can be done as soon as the best fitness dominates for several generations assuming that the solution converged to a local subregion within the search space. Therefore, the selection of the associated population size and the number of generations can be done more efficiently if the entire population (average fitness) is considered where the convergence to a local subregion can be observed closely.

On the other hand, the average fitness of the population is a good indicator of the convergence of the population to local optima. It can be said that when the average fitness of the population is close to the best fitness, the resulting solution is the optimal solution. In other words, when the average fitness of the population does not change with time (from generation to generation), the resulting solution is the optimal solution using the current GA parameters and operators. Therefore, the average fitness is more appropriate to study the convergence of the search and hence the selection of the required population size can be done efficiently.

The diversity of the individuals in the population can be studied easily using the average fitness rather than the best fitness. The variation of the average fitness with generations indicates that no chromosome is dominant and hence the population is diverse. The GA parameters, GA operators, and the population size control the diversity of the population. When the average fitness of several populations is constant (or almost constant), the population becomes dominated by one or few chromosomes and hence the effect of increasing the number of generations is negligible since the GA parameters and operators will not be able to produce new chromosomes. The best fitness gives limited information regarding the diversity of the population and hence it cannot be used to study the performance of the GA parameters and operators. Therefore, the average fitness is more appropriate to study the interaction between the GA parameters and their effect on the GA search process.

Figures 6.36 through 6.46 show the best fitness value in each generation during the genetic algorithms search process, for the pavement section and GA parameters described in Section 6.5.2. The figures show that the use of the best fitness in the sensitivity analysis to select the GA parameters is not appropriate compared to the average fitness values as shown in Figures 6.14 through 6.24. Figures 6.36 through 6.46 show that the change in the best fitness value can be slow when both the crossover and the mutation probabilities are low while the change is high when either the crossover probability or the mutation probability is high. In addition, the best fitness (for sensitivity analysis) relations in Figures 6.36 through 6.46 may suggest the same performance of the genetic parameters on the search space (eg. Figure 6.41) where the best fitness values from different genetic parameter sets have close values. However, the use of the average fitness values in Figures 6.14 through 6.24 show different trends and better conclusions can be obtained.

142

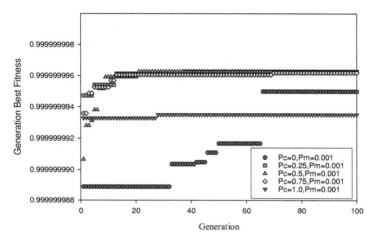

Figure 6.36, Generation best fitness of the backcalculated moduli under a constant mutation rate (P_m=0.001) and variable crossover rates for a large population.

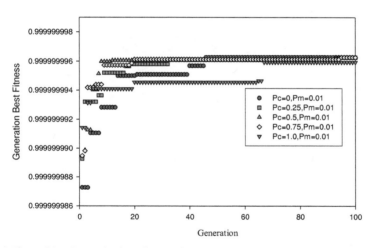

Figure 6.37, Generation best fitness of the backcalculated moduli under a constant mutation rate (P_m=0.01) and variable crossover rates for a large population.

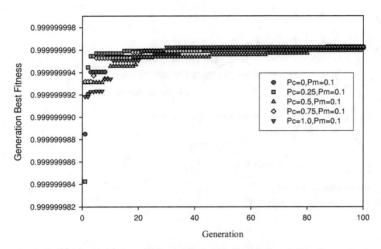

Figure 6.38, Generation best fitness of the backcalculated moduli under a constant mutation rate (P_m=0.1) and variable crossover rates for a large population.

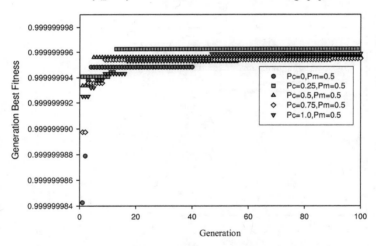

Figure 6.39, Generation best fitness of the backcalculated moduli under a constant mutation rate (P_m=0.5) and variable crossover rates for a large population.

144

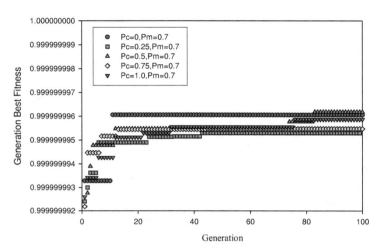

Figure 6.40, Generation best fitness of the backcalculated moduli under a constant mutation rate (P_m=0.7) and variable crossover rates for a large population.

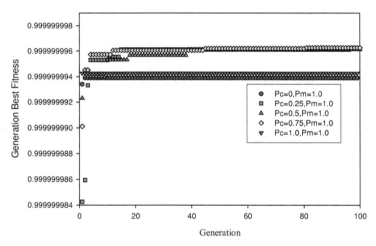

Figure 6.41, Generation best fitness of the backcalculated moduli under a constant mutation rate (P_m=1.0) and variable crossover rates for a large population.

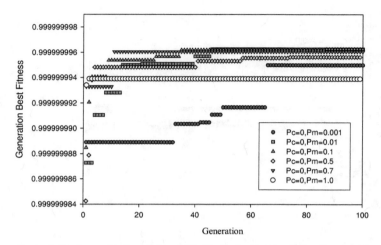

Figure 6.42, Generation best fitness of the backcalculated moduli under variable mutation rates and a constant crossover rate (P_c=0.0) for a large population.

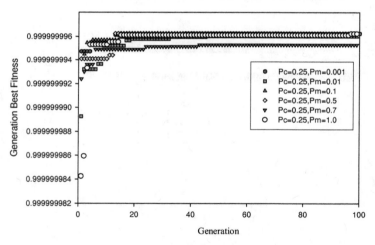

Figure 6.43, Generation best fitness of the backcalculated moduli under variable mutation rates and a constant crossover rate (P_c=0.25) for a large population.

146

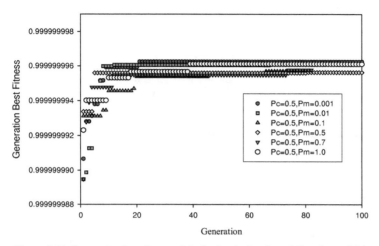

Figure 6.44, Generation best fitness of the backcalculated moduli under variable mutation rates and a constant crossover rate (P_c=0.5) for a large population.

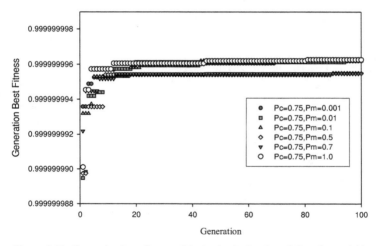

Figure 6.45, Generation best fitness of the backcalculated moduli under variable mutation rates and a constant crossover rate (P_c=0.75) for a large population.

147

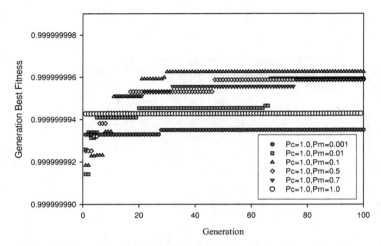

Figure 6.46, Generation best fitness of the backcalculated moduli under variable mutation rates and a constant crossover rate (P_c=1.0) for a large population.

6.6 Conclusions

The backcalculation of the moduli of a pavement system is a very complex process since no closed solution is available or possible for the backcalculation process. In addition, multimodal search spaces can be very complex and may show premature convergence. Therefore, the selection of the genetic operators is very important since poor operators can lead to a premature convergence to a local optimum rather than a global optimum as a result of schema deceptiveness.

It was shown that achieving the same fitness value using different sets of the genetic operators and parameters is possible. However, defining a set of parameters to be used with all GA optimizations showed to be difficult and far from being practical.

The proposed sensitivity analysis showed a more practical and easy tool to investigate the effects of the genetic parameters and operators on the performance of the genetic algorithm and search space. In addition the sensitivity analysis is more appropriate than creating hundreds of trial and error runs based on sets of the genetic operators and parameters. For example, in the trial and error method reported by Reddy et al. (2004) approximately 1200 runs were used to determine the optimal genetic operators and parameters. Using the proposed sensitivity analysis in this chapter, the GA operators and parameters can be determined using 30 runs only.

In addition, the selection of the pavement moduli ranges is as important as selecting the GA operators and parameters. It is recommended that typical values of material moduli be used in the GA backcalculation procedure to avoid local optimal solutions.

CHAPTER VII

FACTORS AFFECTING MODULI BACKCALCULATION

7.1 Introduction

Sensitivity analysis was performed to study the effect of the genetic parameters (population size, total number of generations, mutation rate, and crossover rate). However, other genetic parameters can affect the quality and performance of the optimization process. In the previous pavement studies by Fwa et al. (1997), Kameyama et al. (1997), Reddy et al. (2004), and Tsai et al. (2004), the trials were performed ignoring the effect of the chromosome length, the effect of increasing the number of layers, and the effect of the fitness function. The performance of the genetic algorithms does not depend on the performance of the population size and the mutation and crossover probabilities only. The genetic and evolutionary algorithms work as the natural selection in nature where each operator works on selecting the fittest individuals that can survive the existing conditions and environment. Therefore, relying mainly on one operator or on a partial set of operators can make the selection of the fittest individuals (if the survival of the fittest is retained) a lengthy process. However, in genetic algorithms the development of new methods is still ongoing to better simulate the natural selection and to produce more effective algorithms and operators. Thus, one should consider the interaction between the most effective factors that help in solving the problem and reducing the possibility of early false (premature) convergence.

7.2 Chromosome Length

Chromosomes are formed by encoding the randomly generated solutions (trial pavement moduli) into binary bits (0 and 1). The chromosome length is equal to the total number of digits (length of binary bits) of each encoded solution multiplied by the number of solutions. Therefore, mutation and crossover rates can have different effects on the chromosome length depending on the length of the chromosome. For example, a 011011 chromosome (length=6) will have a different response to mutation and crossover compared to a 0111011101 chromosome

150

(length=10) when encoding two solutions where 3 binary bits were used for each solution in the first chromosome while 5 binary bits were used for each solution in the second chromosome.

The mutation will be carried out by randomly flipping the binary bits (1 becomes 0 and 0 becomes 1), therefore, the probability of flipping the same position in the chromosome is higher when the chromosome length is small (for example 6 versus 10) while the probability of flipping the same position in long chromosomes is lower. This variation in probability will affect the diversity of the chromosomes in the new generation and hence can lead to a premature convergence (false convergence).

On the other hand, the length of the chromosome will affect the number of possible solutions associated with that chromosome according to the schema theorem. According to the theorem, using binary codes, there exist 2^l possible solutions associated with each gene of length l. Therefore, a gene 5 string-long will have $2^5=32$ possibilities while a gene 20 string long will have $2^{20}=1,048,576$ possibilities (trial solutions). Increasing the chromosome length is expected to increase the candidate solutions that the genetic algorithm can explore randomly. Using a complete exhaustive search will be impossible for large chromosome lengths due to time and computational constrains.

The effect of the chromosome length was ignored in all of the pavement backcalculation studies assuming negligible effect. The chromosome length in Fwa et al. (1997), Kameyama et al. (1997), Reddy et al. (2004), and Tsai et al. (2004) was not reported indicating that the studies neglected the effect of the chromosome length assuming negligible or no effect on the performance of the backcalculation process. The ***BackGenetic3D*** program can be used for unlimited number of layers and hence the number of backcalculated parameters will be unlimited based on the user preference and the problem setup. The effect of the chromosome length should be investigated since short chromosomes can affect the performance of the genetic algorithm operators and hence the backcalculated moduli.

7.2.1 Chromosome Length and GA Performance

The effect of the chromosome length is studied using two different problems where the length of the chromosome was varied while the performance of the genetic operators and the backcalculated moduli was observed.

151

The first example involved the study of the performance of a small number of layers. The example of Reddy et al. (2004) was used to investigate the effect of the chromosome length on a 3 layer pavement system. In the second example, the effect of the chromosome length on the performance of the backcalculation process was evaluated using 23 layers. Both examples are discussed in details in the following sections.

The best solution is defined as the solution with the highest fitness and the lowest average moduli error. The used fitness function is:

$$RMSE = \sum_{i=1}^{m}(d_i - D_i)^2 \qquad \text{Eq. (7.1)}$$

$$\text{Maximize } f = \frac{1}{1 + RMSE} \qquad \text{Eq. (7.2)}$$

where $RMSE$ is the root mean square error, m is the number of measuring sensors, d_i is the backcalculated deflection at point i, D_i is the measured deflection at point i, and f is the fitness of the solution based on the root mean square error as estimated in Eq. (7.1).

The "goodness" of the selected parameters using the fitness function in Eq. (7.2) was further assessed using the following equation:

$$RMSE_E = \sqrt{\sum_{i=1}^{3}(E_{ci} - E_{mi})^2} \qquad \text{Eq. (7.3)}$$

where $RMSE_E$ is the root mean square error of the backcalculated elasticity moduli, and E_{ci} and E_{mi} are the backcalculated and the sensitivity pavement system moduli (actual), respectively, for layer i.

7.2.1.1 Genetic Algorithm Performance

The 3-layer pavement system shown in Figure 7.1 was analyzed using the genetic operators shown in Table 7.1 and the moduli ranges shown in Table 7.2. These values were selected arbitrarily. However, the selected genetic operators showed relatively good performance for large populations based on the sensitivity analysis in a previous chapter. The selected parameters may not be the optimal parameters and used here to facilitate studying the GA performance only. The $RMSE_E$ was calculated for six points along the surface of the pavement located at 0,0.3,0.6,0.9,1.2, and 1.5 m from the center of the loading plate.

The data of the 23-layer pavement system is given in Table 7.3 while the genetic operators and the moduli ranges of the backcalculated layers are shown in Tables 7.4 and 6.5, respectively. The 23-layer pavement system represents the modulus variation in pavement due to temperature variation with depth. The arbitrarily selected genetic operators showed relatively good performance for large populations based on the sensitivity analysis in a previous chapter. The selected parameters may not be the optimal parameters and used here to facilitate studying the GA performance only. The radius of the loaded plate was 0.1105 m while the applied load was 690kPa. The $RMSE_E$ was calculated for seven points along the surface of the pavement located at 0,0.203,0.305,0.457,0.610, 0.914, and 1.524 m from the center of the loading plate.

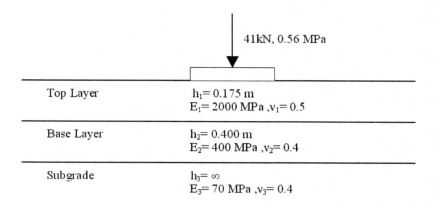

Figure 7.1, The 3-layer pavement system used in the chromosome length analysis, after Reddy et *al.* (2004).

Table 7.1, Genetic parameters used in the backcalculation process (3 layers).

Parameter	Value
Population size	100
Number of generations (G)	100
Probability of crossover (P_c)	0.75
Probability of mutation (P_m)	0.001

Table 7.2, Moduli ranges for the backcalculation process (3 layers).

Layer	Moduli Range (MPa)
Top Layer	1200-2800
Base Layer	240-560
Subgrade Layer	40-100

154

Table 7.3, Data of the 23-layer pavement example.

Description	Layer Number	Thickness (m)	Modulus (MPa)	Poisson's ratio
Asphalt Concrete	1	0.0075	5058	0.30
	2	0.0075	4846	0.30
	3	0.0075	4642	0.30
	4	0.0075	4446	0.30
	5	0.0075	4258	0.30
	6	0.0075	4078	0.30
	7	0.0075	3906	0.30
	8	0.0075	3742	0.30
	9	0.0075	3586	0.30
	10	0.0075	3438	0.30
	11	0.0075	3298	0.30
	12	0.0075	3166	0.30
	13	0.0075	3042	0.30
	14	0.0075	2926	0.30
	15	0.0075	2818	0.30
	16	0.0075	2718	0.30
	17	0.0075	2626	0.30
	18	0.0075	2542	0.30
	19	0.0075	2466	0.30
	20	0.0075	2398	0.30
Base Layer	21	0.25	700	0.30
Subbase Layer	22	0.25	300	0.30
Subgrade Layer	23	∞	100	0.30

155

Table 7.4, Genetic parameters used in the backcalculation process (23 layers).

Parameter	23 Layers
Population size	100
Number of generations (G)	10
Probability of crossover (P_c)	0.75
Probability of mutation (P_m)	0.001

Table 7.5, Moduli ranges for the backcalculation process (23 layers).

Description	Layer Number	Moduli range (MPa)
Asphalt Concrete	1	2000-8000
	2	1900-7800
	3	1900-7400
	4	1800-7100
	5	1700-6800
	6	1600-6500
	7	1600-6300
	8	1500-5600
	9	1400-5700
	10	1400-5500
	11	1300-5300
	12	1300-5000
	13	1200-5000
	14	1200-5000
	15	1100-4500
	16	1100-4300
	17	1100-4200
	18	1000-4100
	19	1000-4000
	20	1000-4000
Base Layer	21	280-1120
Subbase Layer	22	120-480
Subgrade Layer	23	40-160

The 3-layer pavement system was studied using five different chromosome lengths. The assumed chromosome lengths were 18,30,45,60, and 90 binary bits corresponding to parameter lengths of 6,10,15,20, and 30 binary bits. Therefore, a wide range of chromosome lengths was covered in the analysis to extensively study the effect of the chromosome length on the backcalculation process.

Figure 7.2 shows the effect of the chromosome length and the relation between the number of generations, the generation average fitness, and the moduli root mean square error ($RMSE_E$) for the 3-layer example. Figure 7.2a indicates that there is a negligible effect of the chromosome length on the average fitness of the population when the number of generations is larger than 50 (corresponding to more than 5000 evaluations). However, the figure indicates that the required number of generations to achieve high fitness values decreases as the length of the chromosome increases. This performance indicates that the genetic operators perform better when the chromosome length is larger than 30 binary bits (a trial modulus of 10-binary bits long) and the number of generations is less than 50. In addition, for large number of generations (larger than 50) the chromosome length effect on the fitness value is negligible.

On the other hand, Figure 7.2b shows the effect of the chromosome length on the $RMSE_E$ of the moduli. As shown in the figure, the best performance (lowest $RMSE_E$) was achieved when the chromosome length was more than 30 binary bits and the number of generations is larger than 50.

The 23-layer pavement system was studied using five different chromosome lengths. The assumed chromosome lengths were 46, 230,345,690, and 1150 binary bits corresponding to encoded solution lengths of 2,10,15,30, and 50 binary bits, respectively.

Figure 7.3 shows the effect of the chromosome length and the relation between the number of generation, generation average fitness, and the root mean square error ($RMSE_E$) for the 23-layer example. The figure shows that the use of the fitness function as the only measure of the quality and accuracy of the backcalculated moduli can be misleading. In Figure 7.3a, the fitness of the backcalculated displacements was the highest when the chromosome length was 46 while the corresponding $RMSE_E$ value was the highest (233%) in Figure 7.3b indicating a poor performance. On the other hand, for chromosome lengths larger than 46, the fitness values increased as the chromosome lengths increased (Figure 7.3a) while the $RMSE_E$ values decreased

158

as the chromosome lengths increased (Figure 7.3b). The effect of the chromosome length was negligible when the length was larger than 345 binary bits corresponding to an encoded solution length of 15 binary bits.

In addition, Figure 7.3a shows that when the chromosome length is 46, the algorithm converges to a local optimum very quickly. This can be attributed to the fact that with a gene length of 2, the corresponding possibilities will be $2^2=4$, making the search space relatively small and inappropriate for a problem of 23 unknown moduli. Moreover, the crossover and mutation processes were not able to provide enough diversity to avoid the premature convergence. On the other hand, Figure 7.3 shows that it is possible to achieve high fitness values while the backcalculated moduli are far from the actual ones.

It can be concluded that the chromosome length affects the performance of the backcalculation of the moduli when the number of layers is either small (for example 3 layers) or large (for example 23 layers). Based on the above findings, it is recommended that each trial solution be encoded using 15 binary bits or more to increase the accuracy of the backcalculated moduli by reducing the $RMSE_E$ values. Therefore, the chromosome length when backcalculating the moduli should be larger than $15n$ where n is the total number of the unknown moduli or layers.

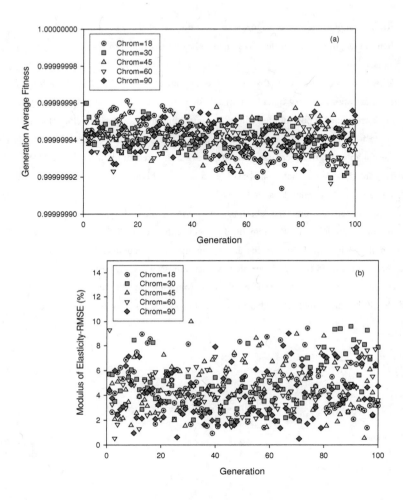

Figure 7.2, Generation average fitness (a) and RMSE of the backcalculated moduli (b) for the 3-layer example.

160

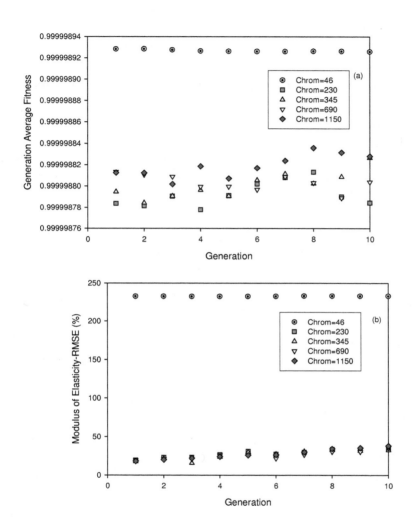

Figure 7.3, Generation average fitness (a) and RMSE of the backcalculated moduli (b) for the 23-layer example.

161

7.3 Importance of The Fitness Function

The genetic operators in genetic algorithms mimic the natural selection in nature by selecting the fittest individuals in the population and by transferring the selected individuals from one generation to the next. The genetic operators work as hill climbers in the local optima domain within the global search domain.

The fitness function in genetic algorithms directs the search process toward the location of the global optimum in the search domain. Therefore, if the used fitness function is not appropriate for the optimization of the problem, the direction of the search will deviate from the right direction and hence the global optima will not be found due to a premature convergence toward a local optimum. The fitness function should be assigned properly to ensure effective optimization of the search space.

In multimodal optimization problems, the selection of the fitness function is vital since the complexity of the search domain can cause a premature convergence toward one of the local optima. When the premature convergence occurs, the genetic operators work on the local search domain (subdomain of the global search domain) and hence the best optimum in that domain is found. Due to the stochastic nature of the genetic algorithms, the search for the global optimum can be a matter of luck if the selection function does not have the ability to properly direct the hard work of the genetic operators to find the optimum.

In addition, the premature convergence is characterized by a remarkable similarity of the individuals indicating low diversity of the population (Minqiang and Jisong, 2001). When the similarity of the individuals in the population increases, the ability of the genetic operators to create better individuals decreases, reducing the order of the created schema. High mutation probabilities have been used to improve the genetic algorithm performance when dealing with complex multimodal optimizations but with little improvement (Minqiang and Jisong, 2001). It was shown in a previous chapter that the use of high mutation rates in the backcalculation of the moduli can be inappropriate sometimes appropriate, since the destruction of the schemata in the backcalculation process increases by increasing the mutation probability.

162

7.3.1 Search and Selection Mechanisms

In addition to the directional effect of the fitness functions, they are used to measure the quality of the selected individuals (chromosomes) and hence they provide feedback to the genetic algorithm. The feedback is then used by the selection process to create the next generation of chromosomes. Tournament selection is the most common method of selection in genetic algorithms. In the tournament selection a number of individuals is selected then the individual with the best fitness is selected based on its fitness value provided by the fitness function. Tournament selection (implemented in **BackGenetic3D**) is performed on all individuals that are involved in the reproduction process including the crossover and mutation.

The genetic operators create new individuals (children) with better characteristics than the previous ones (parents) with some inherited characteristics or totally new characteristics based on the operators' values and the interaction between the operators. However, the effect of the new individuals (chromosomes) can be minimal on the fitness values as measured by the fitness function due to the nature of the fitness landscape. In smooth fitness landscapes, small and large changes to the chromosomes cause in turn small and large changes, respectively, to the associated fitness values (*strong causality*) indicating that there is a correlation between the change in the chromosome and the change in the fitness. Strong causality exists mainly in smooth landscapes. *Weak causality* indicates cases where changes in the chromosome cause changes in the fitness with little or no correlation between both changes (Kauffman, 1993). In some search domains the landscape is rough (not smooth) and hence the possibility of converging to a local subdomain with a local optimum is more likely. The selection of the fitness function largely affects the values associated with the chromosomes and hence affects the rate of change of the fitness. Therefore, it is more appropriate to select a fitness function characterized by strong casualty since small change rates are more desirable (Rosca and Ballard, 1995).

In order to understand the characteristics of the search landscape of the fitness and the search landscape of the associated $RMSE_E$ errors, Figures 7.4 and 7.5 are drawn for large and small population sizes based on the results of the 3-layer example (original ranges of moduli). The figures show that the fitness landscape is smooth regardless of the population size (Figures 7.4a and 7.5a). On the other hand, Figure 7.4b show smooth search landscapes for the $RMSE_E$ regardless of the mutation probability. However, Figure 7.5a shows that the search landscape for the $RMSE_E$ can be smooth or rough at different generation numbers indicating different sensitivities to the change in the fitness values. Since the only information that we have in the real moduli backcalculation problems are the displacements data and the feedback from the fitness function, the need for smooth landscapes for both the fitness and the associated $RMSE_E$ errors is vital. Therefore, large populations are favorable when backcalculating the pavement moduli since the quality of the feedback by the fitness function can somehow be used to measure the quality of the backcalculated moduli.

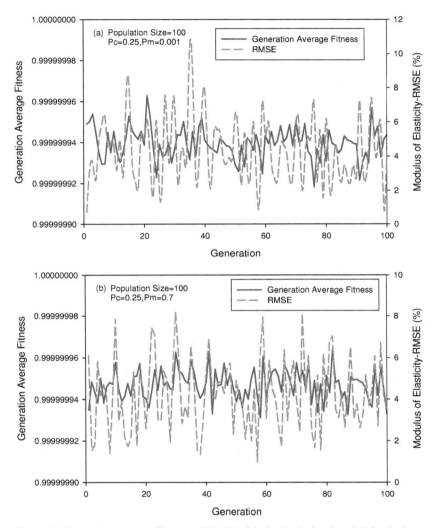

Figure 7.4, Generation average fitness and RMSE of the backcalculated moduli for the large population in the 3-layer example using (a) P_c=0.25 and P_m=0.001, and (b) P_c=0.25 and P_m=0.7.

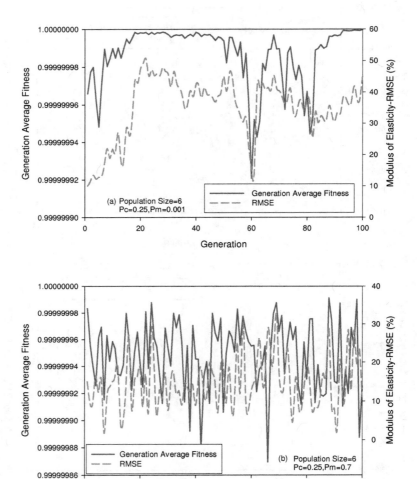

Figure 7.5, Generation average fitness and RMSE of the backcalculated moduli for the small population in the 3-layer example using (a) P_c=0.25 and P_m=0.001, and (b) P_c=0.25 and P_m=0.7.

The selection of the genetic operators and the fitness function is a problem that has been introduced as the exploration-exploitation trade-off problem (Epstein, 1967; Holland, 1992). In genetic algorithms, the fitness function and the genetic operators perform the exploration and exploitation, respectively. Genetic algorithms perform best when the search space exploration initially is performed by weakly causal changes, while the exploitation is later performed by strongly causal changes (Rosca and Ballard, 1995). However, weak causality is more common in genetic algorithm problems.

7.3.2 Least Square Error

The least square error has been used as the fitness function in all of the moduli backcalculation studies using genetic algorithms (Fwa et al., 1997 ; Kameyama et al., 1997 ; Reddy et al., 2004 ; Tsai et al., 2004). The fitness functions used in all studies are shown below:

Fwa et al. (1997):   Minimize $RMSE = \sqrt{\dfrac{1}{m} \sum\limits_{i=1}^{m} \left(\dfrac{d_i - D_i}{D_i} \right)^2}$ Eq. (7.4)

Kameyama et al. (1997): Minimize $RMSE_k = \sum\limits_{i=1}^{m} \sqrt{w_i (d_{ik} - D_{ik})^2}$ Eq. (7.5)

Reddy et al. (2004):   Minimize $RMSE = \sum\limits_{i=1}^{m} (d_i - D_i)^2$ Eq. (7.6)

Tsai et al. (2004):   Minimize $RMSE = \sum\limits_{i=1}^{m} (d_i - D_i)^2$ Eq. (7.7)

where $RMSE$ is the root mean square error, m is the number of measuring sensors, d_i is the backcalculated deflection at point i, D_i is the measured deflection at point i, w_i is the weight factor at point i, and k is the chromosome number.

As can be seen, all of the above functions used the least square errors as a measure of the quality of the backcalculated displacements and provided feedback to the genetic algorithm to direct the search.

The principal idea behind the selection of the least square error is the minimization of the associated errors with each chromosome. When the least square error is minimized, the error between the measured deflectionts from the FWD test is minimal, and hence it is commonly assumed that the difference between the actual moduli and the backcalculated moduli is minimal as well. However, this assumption was shown to be not true since the displacement basin measured in the FWD test can be produced using more than one combination of layers moduli and hence the solution is not unique. In other words, the same fitness value of the chromosome (backcalculated moduli) can be obtained using different moduli. Therefore, it is important to understand the mechanism of the fitness function to determine the effectiveness of the feedback process.

7.3.3 Mathematical Background

In the least square error method a straight line is used to fit between a set of paired observations (Chapra and Canale, 1988) such as (x_i, y_i). If a straight line is assumed to connect between the paired observations, the straight line can be written as:

$$y = a_0 + a_1 x + e \qquad \text{Eq. (7.8)}$$

where a_0 and a_1 are the intercept of the straight line with the y axis and the slope of the straight line ,respectively, while e is the residual between the model and the observations which can be written as:

$$e = y - a_0 - a_1 x \qquad \text{Eq. (7.9)}$$

Therefore, to fit a line through the paired observations, the error associated with the fitting should be minimal to obtain the "best fit". The best fit can be achieved by minimizing the residual errors, for all data points (n), which can be written as:

$$\sum_{i=1}^{n} e_i = \sum_{i=1}^{n} (y_i - a_0 - a_1 x) \qquad \text{Eq. (7.10)}$$

However, the above equation can be satisfied with any line (except a perfectly vertical line) that passes a point in the middle of the line given by Eq. 7.8. Therefore, the above criterion is not satisfactory since no unique best fit is available and hence should not be used.

The above criterion can be modified to:

$$\sum_{i=1}^{n} |e_i| = \sum_{i=1}^{n} |y_i - a_0 - a_1 x| \qquad \text{Eq. (7.11)}$$

where the absolute values of the residuals are minimized to obtain the best fit straight line. Again, this criterion will not yield a unique solution since two lines can have the same absolute residual differences but with different slopes and still satisfy the above criterion.

The problem of the criterion presented in Eq. 7.11 can be solved by minimizing the sum of the squares of the residuals:

$$\sum_{i=1}^{n} e_i^2 = \sum_{i=1}^{n} (y_i - a_0 - a_1 x)^2 \qquad \text{Eq. (7.12)}$$

which yields a unique line for a given set of observation points.

The individual residual values computed by Eq. 7.12 are the squares of the distance between the data points and the straight line. In addition, the quality of the best fit can be evaluated by measuring the "standard error of the estimate" which can be computed by:

$$s_{y/x} = \sqrt{\frac{\sum_{i=1}^{n} (y_i - a_0 - a_1 x)^2}{n-2}} \qquad \text{Eq. (7.13)}$$

which measures the error of the predicted value of y using the value of x by quantifying the spread of the data around the regression line. Another measure that quantifies the spread of the data around the mean ($\bar{y}$) is the standard deviation (S_t) given by:

$$S_t = \sqrt{\frac{(y_i - \bar{y})^2}{n-1}} \qquad \text{Eq. (7.14)}$$

The mean can be calculated by $\bar{y} = \dfrac{\sum_{i=i}^{n} y_i}{n}$. However, it is still possible to obtain a standard deviation (S_t) that is close to 1 while the relation between the y and x is not even linear (Chapra and Canale, 1988).

The above discussion shows that the least squares method has some drawbacks that can prohibit the finding of a unique solution. Even if a unique solution is available by the least squares method, the solution will result in the best "match" between the calculated deflection basins from the backcalculated moduli and the measured deflection basin which does not ensure the finding of the "right" moduli of the pavement. It was shown that different elastic moduli can result in the same displacements residual and hence the same least square errors.

169

7.3.4 Selection of The Fitness Function

It was shown that the fitness function has a significant effect on the genetic algorithm performance. Therefore, the fitness function should be selected properly to enhance the genetic algorithm performance by directing the search process to the subdomain of the best solution (optimum solution).

The following functions were identified as possible fitness functions that can be used to explore the search domain:

Case 1:   Maximize $f = \dfrac{1}{1 + RMSE}$ Eq. (7.15)

Case 2:   Maximize $f = \dfrac{1}{100(1 - RMSE) + 1}$ Eq. (7.16)

Case 3:   Maximize $f = \dfrac{1}{1000(1 - RMSE) + 1}$ Eq. (7.17)

Case 4:   Maximize $f = \dfrac{1}{0.1(1 - RMSE) + 1}$ Eq. (7.18)

Case 5:   Maximize $f = \dfrac{1}{0.01(1 - RMSE) + 1}$ Eq. (7.19)

Case 6:   Maximize $f = \dfrac{1}{0.001(1 - RMSE) + 1}$ Eq. (7.20)

Case 7:   Minimize $f = \sqrt{\dfrac{RMSE}{N}}$ Eq. (7.21)

Case 8:   Minimize $f = r^2 = 1 - \dfrac{RMSE}{\sum\limits_{i=1}^{m}(D_i - \overline{D})^2}$ Eq. (7.22)

Case 9:   Minimize $f = \sum\limits_{i=1}^{N}|d_i - D_i|$ Eq. (7.23)

Case 10:   Minimize $f = \sum\limits_{i=1}^{N}(d_{max} - D_i)^2$ Eq. (7.24)

Case 11:   Minimize $f = \sum\limits_{i=1}^{N}(d_i - D_{max})^2$ Eq. (7.25)

Case 12: Minimize $AREA = \dfrac{1}{2D_0}\left[D_0 r_1 + \left(\sum_{i=1}^{N-1} D_i(r_{i+1} - r_i) \right) + D_N(r_N - r_{N-1}) \right]$ Eq. (7.26)

where N is the number of sensors (measured deflection points), d_i is the backcalculated deflection at point i, D_i and $\overline{D}$ are the measured deflections at sensor i and the average deflection of all sensors, respectively, d_{max} and D_{max} are the maximum backcalculated and measured deflections of all sensors, respectively, and r_i is the distance between the center of the loading plate and sensor i. The RMSE in the above equations is given by Eq. (7.6).

Case 1 was proposed by Reddy et al. (2004) to minimize the RMSE error. Cases 2 through 6 have the same formula but with different coefficients (100,1000,0.1,0.01, and 0.001). Those cases (Cases 2 through 6) represent nonlinear fitness amplifications that have been used to increase the sensitivity of the selection process to any variation in the gradient of the fitness surface in cases where the surface is flat (Daridi et al., 2004). Cases 7 and 8 use the correlation coefficient as the objective function. Cases 9 through 11 are different forms of the least square error.

Case 12 uses the AREA parameter as the objective function since it combines the effect of several measured deflections. The AREA parameter was first proposed by Hoffman and Thompson (1981) for interpreting flexible pavement deflection basins. The AREA parameter has a length dimension since it is normalized using one of the measured deflections to remove the effect of the load.

Figure 7.6 shows the effect of the fitness function on the average backcalculated moduli. As the figure shows, the fitness functions are not equal in their effect even though the least square error or a modified form of it was used in all cases except Case 12. The $REMSE_E$ error of one fitness function (eg. Case 6) can be more than twice that from another fitness function (eg. Case 12). This indicates that the exploration mechanism in some cases performed better in guiding the genetic algorithm toward the best average solution while other functions guided the genetic algorithm toward other domains that have local optima.

In addition, the convergence of the fitness functions varied from one function to another. In Case 12 the fitness function converged quickly to the optimal solution while the other functions showed some stochastic behavior in the $REMSE_E$ error. On the other hand, all cases

171

showed an increase in the $REMSE_E$ error with the increase of the generation while Case 12 showed steady $REMSE_E$ error values.

The function represented by Case 12 has shown the best performance among the studied functions. Case 11 showed lower $REMSE_E$ error values at the beginning of the optimization but then showed high $REMSE_E$ error values later on. Moreover, Case 12 showed *strong causality* indicating that a small change rate exists in the search space of the moduli, which is desirable (Rosca and Ballard, 1995). Since the use of the GA does not have any clue about the "actual" moduli, the fitness function represented by Case 12 is recommended over the other tested functions.

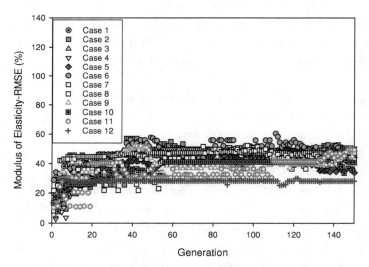

Figure 7.6, Generation's RMSE of the backcalculated moduli for the 3-layer example.

7.4 The Effect of Increasing the Number of Layers

Increasing the number of layers in the pavement system will increase the number of the backcalculated moduli and hence the number of unknowns. In addition, increasing the number of unknowns increases the complexity of the search domain and can affect the performance of the

172

genetic parameters and operators. Therefore, it is vital to understand the effect of increasing the number of unknown moduli on the performance of the backcalculation process to better select more appropriate genetic parameters and hence to design more effective backcalculation algorithms.

In order to study the effect of the number of layers, the 3-layer pavement system shown in Figure 7.1 was modified to a 4-layer pavement system by adding an additional subbase layer below the base layer and above the subgrade layer. The data of the new 4-layer pavement system is shown in Table 7.6 while the used genetic parameters and moduli ranges are shown in Tables 7.7 and 7.8, respectively. The effect of adding the layer was studied using a large population size (population size of 100).

Table 7.6, Data of the 4-layer pavement example.

Description	Layer Number	Thickness (m)	Modulus (MPa)	Poisson's Ratio
Top Layer	1	0.175	2000	0.30
Base Layer	2	0.400	400	0.30
Subbase Layer	3	0.200	200	0.30
Subgrade Layer	4	∞	70	0.30

Table 7.7, Genetic parameters used in the backcalculation process (4 layers).

Parameter	Value
Population size	100
Number of generations (G)	100
Probability of crossover (P_c)	0.75
Probability of mutation (P_m)	0.001

173

Table 7.8, Moduli ranges for the backcalculation process (4 layers).

Layer	Moduli Range (MPa)
Top Layer	1200 - 2800
Base Layer	240 - 560
Subbase Layer	120 - 280
Subgrade Layer	40 - 100

Figure 7.7 shows that using the same genetic operators and parameters that were used in the 3-layer pavement system to backcalculate the unknown pavement moduli in the 4-layer pavement system can lead to a better performance. The figure indicates that the complexity of the search space in the 4-layer system can be explored and exploited more effectively even though the used genetic operators and parameters showed a relatively poor performance when they were used to explore and exploit the 3-layer system which is relatively less complex. In addition, no rule of thumb can be drawn from increasing or reducing the number of the unknown moduli on the performance of the genetic backcalculation process. This indicates that no unique set of genetic operators can be used blindly with any pavement system due to the complexity of the search space, the interaction between the genetic operators, and the effect of the genetic operators on the search space.

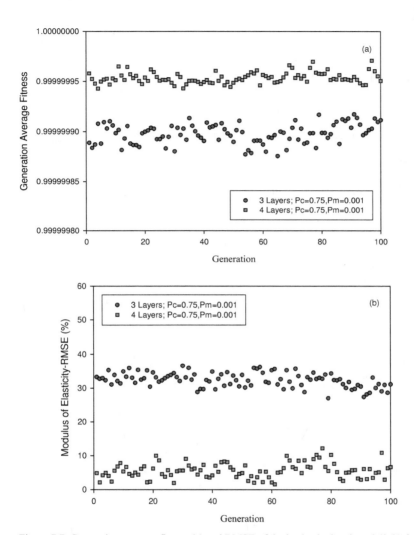

Figure 7.7, Generation average fitness (a) and RMSE of the backcalculated moduli (b) for the 3-layer and 4-layer pavement systems.

175

7.5 The Effect of the Number of Sensors

The effect of the number of sensors on the backcalculated moduli was studied using the 3-layer pavement system and assuming two different objective functions. The first objective function is based on the RMSE error (best-fit case) as given by Eq. (7.15) while the second objective function is based on the AREA method (AREA case) as given by Eq. (7.25).

The relations between the number of generation and the root mean square error of the backcalculated moduli are shown in Figures 7.8 and 7.9 for the best-fit case and the AREA case, respectively. Both methods indicated that increasing the number of sensors will increase the accuracy of the backcalculated moduli. However, the effect of the number of sensors on the accuracy of the backcalculated moduli is clearly evident in the AREA case where the accuracy (lowest RMSE) was the highest when all of the sensors (6 sensors) were included in the backcalculation process and was the lowest (high RMSE) when only one sensor was included.

Figure 7.8 shows the highest accuracy when all of the sensors (6 sensors) were included in the backcalculation process. However, there is no distinct trend between the increase in accuracy and the number of sensors when the number of sensors is less than 6. This indicates that the best-fit case employs a poor objective function that affects the exploitation and exploration for the optimal solution.

On the other hand, Figure 7.9 shows the sensitivity of the backcalculation process to the number of sensors. The figure indicates that increasing the number of sensors can greatly affect the performance of the backcalculation process and the interaction between the genetic operators and parameters to explore and exploit the search space. In other words, the complexity of the search space is affected by the number of sensors which in turn affects the accuracy of the backcalculated moduli. In Figure 7.9, doubling the number of sensors (3 sensors to 6 sensors) reduced the error by 50%. However, the figure indicates that there is a threshold for the number of sensors after which the reduction in error can be negligible (5 sensors and 6 sensors).

It can be said that increasing the number of sensors increases the accuracy of the backcalculated moduli. In addition, the selection of the objective function in the backcalculation process affects the optimal number of sensors as well as the accuracy of the backcalculated moduli.

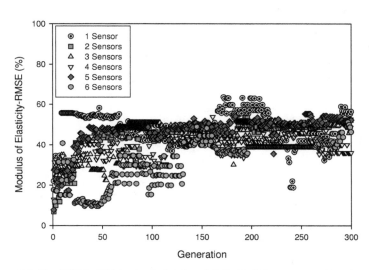

Figure 7.8, The RMSE of the backcalculated moduli for the 3-layer pavement system using different number of sensors (RMSE method).

177

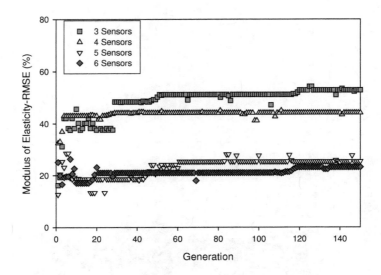

Figure 7.9, The RMSE of the backcalculated moduli for the 3-layer pavement system using different number of sensors (AREA method).

7.6 Conclusions

The backcalculation of the pavement moduli is sensitive to many factors other than the genetic algorithms parameters and operators. It was found that the backcalculation process is sensitive to the chromosome length, the number of layers, the number of sensors, and the objective (fitness) function.

It is evident that increasing the chromosome length would increase the accuracy of the backcalculated moduli since a larger search space would be explored. Increasing the number of layers can either increase or decrease the complexity of the genetic operators work to find the optimal solution. The AREA method showed the best performance when used as the fitness function compared to many other objective functions.

The effect of increasing the number of sensors depends on the used fitness function. However, there is a threshold beyond which the effect of the number of sensors will be negligible.

178

CHAPTER VIII

PARAMETERLESS GENETIC ALGORITHM FOR THE BACKCALCULATION OF

ELASTIC MODULI

8.1 Introduction

Genetic algorithms are very powerful algorithms for the optimization of complex functions and applications. The mechanism of genetic algorithms is similar to that from natural selection theory where a population first is randomly created (candidate solutions), and then mating between the individuals (chromosomes) starts based on the characteristics (fitness value) of the individuals. The mating process produces new individuals that can survive the surrounding environment and conditions and should ensure that the best characteristics that can handle the environment will transfer (will be inherited) to the next generation (offspring).

The reproduction of chromosomes in genetic algorithms is controlled by the genetic operators such as the crossover and the mutation. These operators work together to exploit the search space by providing diversity among the individuals of the generation to allow a wider spatial investigation of the subdomain (within the search space) that was pointed to by the fitness function. However, the interaction between the genetic operators is very complex since each operator has its own effect on the search space and the schema of the newly generated chromosomes. For example, using high mutation probabilities can disturb the schema which in turn will lead to a complete random generation of the chromosomes which in turn will have an adverse effect on the genetic algorithm ability to locate the best solution since less characteristics will be transferred from one generation to the next.

The interaction between the genetic operators will control the population size needed to solve the problem in hand. This can be understood by looking at the genetic operators as "tools" for finding the best solution within a subdomain and hence for convergence. These "tools" can be either highly effective or poorly designed where the time needed to reach the peak or the valley is highly impractical. Therefore, designing the best "tools" will be a challenge since each problem has its unique characteristics and should be considered carefully.

179

The selection of the right genetic operators is not a trivial task. The population size, the search domain, the physical behavior of the optimized problem, and the desired time to convergence all work as constraints to the selection of the genetic operators to be used in practical applications.

8.2 Genetic Algorithms Difficulties

The complexity of the problem, the genetic operators, and the population size are the most common problems when using the genetic algorithms since the interaction between these factors is complex. The content and the context of the genetic algorithm problem and operators are additional unseen factors that affect the problem. The content of a problem is represented by the physical behavior of the problem while the context is represented by the way that the genetic operators interact with each other (Gustafson, 2004). Therefore, one cannot generalize the selected genetic operators that work with one problem to other problems since each problem has its unique content and context. The content and context of a problem can highly affect the relation between the fitness values and the chromosomes where it can become either sensitive or not sensitive.

The population size of any problem influences both the context and the content of the problem. The genetic operators (context) work on the chromosomes of the population and therefore, increasing the population size will tend to create new individuals that behave differently than those generated in a small population. This can be attributed to the higher diversity among the individuals in large populations compared to small ones. In turn, the fitness values associated with large populations comprise a different search landscape than the ones encountered in small populations. The ability of the genetic algorithms to reach optimal solutions is limited. This limitation arises from the behavior of the genetic operators, the fitness function, and the population size where they can lead to a premature convergence.

The increase of the individuals with the same or close fitness values will reduce the ability of the selection procedure. In the selection procedure individuals are compared based on their fitness, therefore, increasing the individuals with the same fitness will make the selection process more random since no preference is given to any of the selected chromosomes. Hence, increasing the fitness with the reduction of the diversity in the population reduces the

180

effectiveness of the genetic algorithms since the algorithm become more random rather than guided.

Deception is another factor that affect the performance of genetic algorithms (Goldberg, 1987; Goldberg, 2002). Deception occurs when the fitness of two chromosomes (solutions) is the same but their values are different, therefore, applying genetic operators such as crossover produces poor fitness values.

8.3 Diversity in Genetic Algorithms

A population of chromosomes is said to be diverse if the chromosomes consist of genes (solutions) from as many subdomains as possible. Diversity is believed to be a vital requirement to avoid premature convergence in genetic algorithms (Ryan, 1994; McKay, 2000). Increasing the diversity can undermine the search process and hence no optimal solution can be found. On the other hand, decreasing the diversity can either slow down the convergence to optimal solutions or result in premature convergence.

However, preserving diversity is not trivial since each run can start with a diverse population over the search domain, but during the run, the population tends to "propagate" toward a subdomain and hence will tend to loose diversity. The initial parent pool is diverse since all individuals are generated randomly. The algorithm starts to converge when the fitness of the individuals cannot be improved indicating the arrival at the optimal solution for the given conditions.

As the genetic algorithm loses diversity the search tends to work more on localized subdomains with close neighborhoods only and loses the ability to work on larger subdomains. Therefore, the search turns from a guided random genetic search to a stochastic hill-climbing search. Hill-climbing is a local search method that moves in a direction related to the local gradient (Goldberg, 1989). In genetic algorithms, hill- climbing can be obtained if the crossover probability is equal to zero while the mutation probability is very small. This indicates that when diversity is very minimal, the only source of diversity would be mutation since crossover will not be able to provide diverse individuals.

181

In addition, diversity is affected by the selection pressure (Whitley, 1989). Increasing the selection pressure increases the convergence ability of the search while decreasing the selection pressure reduces the ability of the search to converge. When the selection pressure increases, the search focuses more on the individuals with the highest fitness and therefore the diversity decreases. In addition, tuning the genetic parameters for a certain problem is in fact an auxiliary way of tuning the selection pressure. On the other hand, increasing the population increases the schemata involved in the search and hence reduces the selection pressure and increases the diversity. The selection procedure (tournament selection, roulette wheel,...etc.) affects the selection pressure. For example, in the roulette wheel selection method individuals with high fitness can dominate quickly leading to a premature convergence.

Controlling the diversity of the population can be achieved by controlling the genetic operators. Controlling diversity showed problematic issues such as knowing the level of diversity needed to solve the problem before the start of the run. The interaction between the exploration and the exploitation processes in genetic algorithms impose a random effect on the generated individuals that make the prediction of the needed level of diversity difficult.

8.4 Interaction Between Genetic Operators

Crossover and mutation are the main genetic operators that are commonly used in genetic algorithms. These operators are applied after calculating the fitness of each chromosome in the population. Then, the operators work to increase the number of the fittest individuals in the population. The schema theorem cannot be used to explain the mechanisms of the search strategy of genetic algorithm (Mühlenbein, 1991).

The crossover is one of the important genetic operators. It works on chromosomes by joining parts of chromosomes together to form new chromosomes. The crossover probability is high in many of the applications in genetic algorithms. On the other hand, mutation works on chromosomes by flipping the binary strings at random locations to its opposite binary value.

The work of De Jong (1975) was the first published attempt to tune the genetic parameters to arrive at a set that can be used for other problems. In his work different functions were tested to cover a wide range of possible complexity in function characteristics. The genetic algorithm components and operators included the single-point crossover, mutation, and the roulette wheel

182

selection method. Many trial runs for different sets of genetic operators/parameters were carried out to better understand the behavior of the genetic algorithms including the crossover probability, the mutation probability, and the population size. The main findings in De Jong's (1975) work included:

1- Mutation is needed to replace the lost chromosomes during the generation of new chromosomes. The mutation probability should be selected carefully and should be low since high mutation rates can result in a random search strategy rather than a genetic algorithm search.

2- The following parameters can be used to optimize a wide range of problems: crossover probability=0.6, mutation probability=0.001, and population size=50 to 100.

Although De Jong's work break the ground for future investigations and research, the recommended genetic parameters cannot be used for all problems as done by many researchers. For example, the use of roulette wheel will impose a different selection pressure than the selection pressure by the tournament selection method. In addition, the search landscape is problem dependent.

Schaffer et al. (1989) expanded the work of De Jong (1975) by adding more functions to those studied by De Hong. They investigated the performance of the functions using different set of genetic operators including crossover probabilities ranging between 0.05 and 0.95, mutation probabilities ranging between 0.001 and 0.1. In addition, Schaffer et al. (1989) used six population sizes including 10,20,30,50,100, and 200. The main findings in their work included:

1-High mutation probabilities are recommended for small populations while small mutation probabilities are recommended for large populations.

2-Mutation can be used alone to solve genetic algorithm problems.

3-The following parameters can be used to optimize similar functions: crossover probability=0.75 to 0.95, mutation probability=0.005 to 0.01, and population size=20 to 30.

Schaffer et al. (1989) found, empirically, that an optimal parameter setting for the studied functions can be achieved using:

$$\ln N + 0.93 \ln m + 0.45 \ln n = 0.56 \qquad \text{Eq. (8.1)}$$

where N is the size of the population, m is the mutation probability, and n is the length of the chromosome. This relation was proved to be incorrect (Mühlenbein, 1992). However, the above study was still limited to a certain number of functions and cannot be extended to all problems. In addition, Schaffer et *al.* (1989) recognized that the used functions are insensitive to crossover and sensitive to mutation and hence more complex functions should be studied.

The effect of the mutation probability on the performance of the genetic algorithms was studied by Mühlenbein (1992) and Bäck (1993). Both studies indicated that the optimal mutation probability for unimodal functions is $1/l$ where l is the length of the chromosome. In addition, Bäck (1993) indicated that for multimodal functions, mutation probabilities equal to $1/l$ were appropriate.

Ochoa et *al.* (1999) showed that without recombination and using adaptive mutation, the mutation probability decreases with time and hence high mutation rates should be used at the beginning of the run. In addition, they showed that with crossover, the optimal solution can be achieved with constant low mutation probabilities.

8.5 Automation of GA

The main problem with genetic algorithms is the selection of the required genetic parameters (genetic operators, population size, and number of generations). The selection of genetic parameters depends on the user experience, published literature, and the trial and error technique. These make the genetic algorithms less appealing to practitioners since no general guidelines are available to guide the users. Therefore, tuning the genetic algorithms parameters is the main challenge that yet has to be solved to guarantee the acceptance of the method by a wide spectrum of practitioners.

The value of the genetic operators can be changed using different methods that can be grouped into three categories (Eiben et *al.*, 1999):

1-Deterministic Parameter Control Methods: In these methods the operators are changed based on deterministic rules without taking into consideration the search progress. The controlling parameter for the modification of the genetic operator is the generation number.

2-Adaptive Parameter Control Methods: In these methods, the direction and/or magnitude of the change of the genetic operator is determined based on the search progress.

3-Self-adaptive Parameter Control Methods: In these methods, the genetic operators are encoded into the chromosomes and hence undergo changes as the search progress.

184

Some of the relationships that have been suggested in literature using deterministic parameter control are presented below.

Fogarty and Terence (1989) proposed the following relation to change the mutation probability (P_m) based on the generation number (t):

$$P_m = \frac{1}{240} + \frac{0.11375}{2^t}$$
Eq. (8.2)

Hesser and Manner (1991) proposed the following relation to change the mutation probability (P_m) based on the generation number (t):

$$P_m = \sqrt{\frac{\alpha}{\beta}} \times \frac{e^{-\frac{\gamma t}{2}}}{\lambda\sqrt{L}}$$
Eq. (8.3)

where λ is the population size, t is the generation number, L is the chromosome length in bits, and (α, β, and γ) are constants.

Bäck and Schutz (1996) proposed the following relation for mutation probability (P_m) based on the generation number (t):

$$P_m = \left(2 + \frac{l-2}{T-1}t\right)^{-1}$$
Eq. (8.4)

where T is the maximum allowed number of generation. This relation constrains the change of the mutation between 0.5 at $t=0$ to $1/L$ at $t=T$.

Figure 8.1 shows the relation between the generation number and the mutation probability as suggested by Fogarty and Terence (1989) and Bäck and Schutz (1996). As can be seen, Fogarty and Terence (1989) method resulted in a mutation probabilities of 0.061042 at generation $t=1$ and a mutation probability of approximately 0.004167 for all generations at and beyond $t=18$. Therefore Fogarty and Terence (1989) suggested relation gives limited number of small mutation probabilities throughout the generations and hence will have effect at the beginning of the GA run while it will be constant after generation 18. On the other hand, Bäck and Schutz (1996) suggested relation showed higher mutation probabilities than those by Fogarty and Terence (1989). In addition, Bäck and Schutz (1996) mutation probabilities were higher for short chromosomes ($l=8$) compared to long chromosomes ($l=16$ or 32). Therefore, it is expected to see higher diversity in the population when the chromosome length is relatively short. Both

185

mutation relations showed that the mutation probability relation should be selected carefully since each problem may respond differently to the rate of change in the mutation probability.

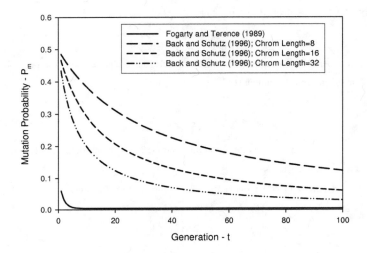

Figure 8.1, Mutation probability as a function of generation number.

The resulted genetic parameters from the manual tuning of parameters using the trial and error technique showed some drawbacks (Eiben et al., 1999). The genetic parameters are independent and hence it is impossible to try all of the different parameter combinations systematically. In addition, the trial and error process is time consuming since a very large number of combinations should be used. On the other hand, the selected parameter combinations will aid in finding the optimal parameter set among the searched possibilities and the optimal parameter set may not be found.

The genetic parameters can be constant during certain stages of the search while they can be adaptive during other stages (Syswerda, 1991; Bäck, 1992; Eiben et al., 1999). For example, large mutation probabilities are needed during the initial stages of the search to aid in the

186

exploration of the search space while small mutation probabilities are needed during the final stages to aid in finding the optimal solution.

Adaptive parameter control was first introduced by Rechenberg (1973). In his work he used a variable mutation probability that was dependent on the rate of successful mutations in a population. Julstrom (1995) changed the value of the crossover and mutation probabilities based on their contribution to the fitness of the new generation.

Srinivas and Patniak (1994) proposed assigning mutation and crossover probabilities for each chromosome based on its fitness and the fitness of the population. Another way of self-adapting was first introduced by Schwefel (1997). He proposed a new method where the encoded chromosome carried also its mutation probability which was subjected to crossover and mutation with the chromosome. Bäck et al. (2000) extended Schwefel (1997) work by encoding the mutation and crossover probabilities with the chromosome. The mutation and crossover probabilities were subjected to evolution during the progress of the run.

8.5.1 Adaptive Population Sizing

The population size is a very important parameter in genetic algorithms. If the population size is underestimated, the algorithm will never converge or will converge to a local optimum since the representation of the search space in the genetic pool is less than what is required, and therefore, regardless of the genetic operators or the selection method, the newly generated schemata will not provide the required diversity in the population. On the other hand, if the population size is overestimated, the algorithm will spend significant time searching for the optimal solution. It is typical to expect larger population sizes in problems of high complexity such as multimodal problems. In other words, increasing the possibility of convergence due to the large number of local optima increases the need for larger population size. Therefore, the selection of the appropriate size is as important as selecting the genetic operators or the selection strategy.

However, selecting the right population size for practical problems is very difficult since the search landscape is not known in advance. A common way of handling this problem is using different population sizes to figure out the effect of the population size on the performance of the algorithm. This trial and error method adds more complexity to the problem of defining the appropriate parameters and operators for the genetic algorithm since different parameter/operator sets can produce different effects. Therefore, changing the size of the population can add an additional variable that need to be optimized.

Different adaptive population sizing methods have been proposed in literature. The following discusses some of the proposed methods.

8.5.1.1 Population Sizing using Schema Variances

This method was first proposed by Smith (1993) and Smith and Smuda (1995) based on the work of Goldberg et al. (1993). In this method the size of the population was adjusted as the search for the optimal solution progressed. They proposed an algorithm that sizes the population based on the expected selection loss and a specified target accuracy should be provided by the user. The expected selection loss between two competing schemata is the probability that the lower fit schema gets selected weighted by the fitness difference between them.

8.5.1.2 Population Sizing using Life Time

Arabas et al. (1994) created a new variable called the *remaining life time* (RLT), which was assigned to each chromosome. The RLT of each chromosome was updated at the beginning of each generation using a bi-linear formula depending on whether the fitness of the chromosome was less than the average fitness of the population or not. Chromosomes with RLT values close to zero are eliminated from the generation. A reproduction factor was used in which part of the population was allowed to reproduce. Different reproduction factors showed different effects on the performance of the genetic algorithm. The method requires an initial population size to be provided by the user.

8.5.1.3 Population Sizing using Competing Subpopulation

Schlierkamp and Muhlenbein (1996) proposed running more than one population at the same time. After several generations, the run with the best fitness is increased while the other generations are reduced. Hinterding et *al*. (1996) proposed running three populations simultaneously with initial size ratios of 1:2:4. The fitness values of the three populations are

188

compared after several runs and the size of the best run is doubled, the size of the worst run is halved, and the size of the run in between is maintained.

8.5.1.4 Population Sizing Using Parameterless GA (PGA)

This method was proposed by Harik and Lobo (1999) and Lobo (2000) assuming that the solution quality increases as the population size increases. They further assumed that for two identical genetic algorithms with two different population sizes (P_1, P_2) where $P_1 < P_2$, if P_1 spends more evaluations (generation multiplied by population size) than P_2 and the average fitness of P_1 is less than P_2, then there is a strong evidence that that the analysis of P_1 is carried out using an undersized population. The algorithm steps are as the following:

1-Run two population sizes simultaneously with $P_i = (1/2) P_{i+1}$, where i is the population number.

2- Run population P_i four times population P_{i+1} (number of generations).

3-Inspect the average fitness of the population after several evaluations. If the average fitness of $P_i < P_{i+1}$ then eliminate P_i and double P_{i+1}.

4-Repeat until the desired fitness is achieved.

According to the schema theorem, the survival probability of a schema can be increased either by increasing the selection rate or lowering the crossover probability (Harik and Lobo, 1999; Lobo, 2000). Both the crossover and the selection rate control the growth of the building blocks from generation to another. In addition, the selection rate controls the amount of bias towards better individuals while the crossover probability controls the amount of recombination or mixing. However, if the selection rate is high the algorithm will pay more attention to the individuals with the highest fitness values while it will pay less attention if the selection rate is too low. An appropriate selection rate should ensure that the growth ratio of building blocks is greater than 1.

The net growth ratio on schema H at generation t is (Harik and Lobo, 1999; Lobo, 2000):

$$\phi(H,t)[1 - \varepsilon(H,t)] \qquad \text{Eq. (8.5)}$$

where $\phi(H,t)$ is the effect of the selection operator, $\varepsilon(H,t)$ is the disruption factor on the schema.

189

The above equation is a simplified form of the schema theorem, which can be written, using the selection rate s and the probability of crossover P_c, as:

$$s(1 - P_c) \qquad \qquad \text{Eq. (8.6)}$$

Harik and Lobo (1999) and Lobo (2000) suggested using $s=4$ and $P_c=0.5$ in the PGA which give a net growth of 2. Mutation probability was ignored for simplicity in the original work by Harik and Lobo (1999) and Lobo (2000). However, Lobo (2000) recommended using a mutation probability equal to the inverse of the chromosome length ($1/l$).

In addition, the PGA takes advantage of the fact that increasing the population size makes the sampling of the building blocks better since the error in decision-making reduces. Population sizing equations proposed in literature (Goldberg et al., 1992; Harik et al., 1997) are difficult to apply in practice. Population sizing equations assume proper mixing of the building blocks. In addition, applying the population sizing equations requires knowing the maximum deception in the problem and the selective advantage of a building block over its most tough competitor (Lobo, 2000).

The selection of an adaptive function is very difficult since many factors are involved during the search process and the interaction between these factors is highly complex. In addition, the adaptive parameters are controlled mainly by the generation number while the actual progress in solving the problem and the performance of the genetic algorithms are ignored (Eiben et al., 1999).

8.6 The Dynamic Parameterless Genetic Algorithm

The optimization of pavement layer moduli by backcalculating the moduli using the FWD data is a complex task since the resulting function is multimodal. Multimodal functions are difficult to optimize due to the high probability of converging to a local optima during the search process. Due to the complexity of the problem, tuning the genetic parameters and operators can be a complex task, as well, which may limit the use of the method by practitioners. The genetic algorithm method is robust and very reliable in finding the global optima if the optimal genetic operators and parameters are used making it more attractive than the available backcalculation

190

procedures. Therefore, the Parameterless Genetic Algorithm method is appealing since none of the parameters needs human tuning.

A new PGA method is proposed. The new PGA method is intended to increase the efficiency of the PGA and to reduce the time associated with the method. The new PGA method is called the Dynamic Parameterless Genetic Algorithm (DPGA). The DPGA is based on Harik and Lobo's (1999) and Lobo's (2000) parameterless GA, and includes the following steps:

1-Run population P_i for a number of generations until the average fitness of two consecutive generations is less than a specified tolerance value or negligible.

2-Run another population (P_{i+1}) where P_{i+1}, is twice P_i.

3-Insert the fittest individuals from run P_i to the parent's pool of population P_{i+1}. The remaining individuals of the parent's pool of population P_{i+1} are generated randomly

4-Inspect the average fitness of the population after several evaluations. If the average fitness of two consecutive generations is less than a specified tolerance value or negligible, stop the run. Otherwise, continue the run.

5-Repeat steps 2 through 3 until the change in the average fitness between two populations (P_i and P_{i+1}) is less than a tolerance value or negligible.

The DPGA is intended to increase the population size of the same run dynamically. By doing so, the diversity of the population is increased dynamically without destructing the schema of the individuals in the population. In the DPGA algorithm, the mutation rate is implicitly employed during the backcalculation process by transferring the best individuals from one run to the other (from P_i to P_{i+1}) while adding randomly the same number of individuals to the next run (to P_{i+1}). Therefore, the mutation rate is random but at the same time is the highest at the beginning of the run (as desired) and decreases as the run progress. In addition, the DPGA implements the niching concept implicitly by giving focus to other than one region within the search space when the highest fitness is focused at one region.

The effect of the PGA and the DPGA algorithms on the performance of the backcalculation process was evaluated using the 23-layer pavement system. The example details were discussed in a previous chapter. The best solution is defined as the solution with the highest fitness and the lowest average moduli error. The used fitness function is:

$$RMSE = \sum_{i=1}^{m} (d_i - D_i)^2 \qquad \text{Eq. (8.7)}$$

191

$$\text{Maximize } f = \frac{1}{1 + RMSE} \qquad\qquad \text{Eq. (8.8)}$$

where $RMSE$ is the root mean square error, m is the number of measuring sensors, d_i is the backcalculated deflection at point i, D_i is the measured deflection at point i, and f is the fitness of the solution based on the root mean square error as estimated by Eq. (8.7).

The "goodness" of the selected parameters using the fitness function in Eq. (8.8) was further assessed using the following equation:

$$RMSE_E = \sqrt{\sum_{i=1}^{3}\left(E_{ci} - E_{mi}\right)^2} \qquad\qquad \text{Eq. (8.9)}$$

where $RMSE_E$ is the root mean square error of the backcalculated elasticity moduli, and E_{ci} and E_{mi} are the backcalculated and the sensitivity pavement system moduli (actual), respectively, for layer i.

Figures 8.2 and 8.3 show the relation between the average fitness of the generation and the RMSE of the moduli with the number of generation for the PGA and the PDGA algorithms, respectively. In the PDGA runs, the case with a population of 64 was used as the starting population and was similar to that in the PGA case. The figures show that the PDGA is better than the PGA since higher fitness values and lower moduli RMSE errors were obtained. In addition, Figure 8.3 shows that the PDGA algorithm is better in exploring and exploiting the search space than the PGA algorithm (Figure 8.2). The PDGA algorithm, in this example, showed RMSE errors approximately 40% lower than those obtained using the PGA since the exploitation and the exploration with mutation and niching were all employed correctly to the search space.

It is evident that increasing the population size "suddenly" increases the ability of the genetic algorithms to explore and exploit other search domains while either focusing more or ignoring the current domain. In the PGA, ignoring the results of one population size and moving to the next population size can duplicate the effort by the genetic algorithms in searching the same subdomains which can be time consuming.

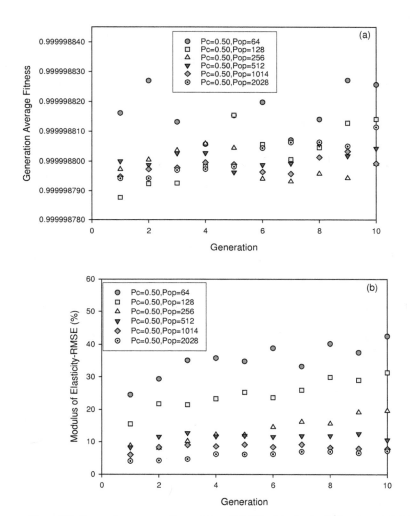

Figure 8.2, Generation average fitness (a) and RMSE of the backcalculated moduli (b) for the 23-layer example using the PGA method.

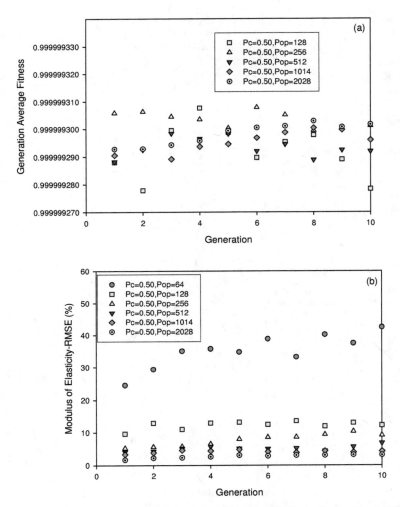

Figure 8.3, Generation average fitness (a) and RMSE of the backcalculated moduli (b) for the 23-layer example using the DPGA method.

8.7 Conclusions

The optimization of the genetic parameters is challenging and can affect the results of the backcalculation process. The results of one optimization problem cannot be used with other problems due to the uniqueness of each problem and the complexity of the search space. On the other hand, the available adaptive methods are developed based on the success of solving certain problems ranging between simple to complex. However, multimodal problems can be challenging and none of the available adaptive method can deal with the complexity of such search spaces.

The automation of the genetic parameters can reduce the effort of the user and can increase the time efficiency in finding accurate results. The automation of the genetic parameters can be carried out using population sizing utilizing the PGA algorithm or the new PDGA algorithm. The PDGA algorithm showed better performance dealing with complex multimodal search spaces such as those in the pavement backcalculation problem. At the beginning of each PDGA set of generations, half of the population is inserted from the previous run while the second half is created randomly. The mutation probability at the beginning of each set of generations is very high and reduces implicitly as the backcalculation progresses. In addition, niching and is implicitly satisfied by giving different than one subdomain some focus through the random distribution of half of the population. Transferring the "fittest" individuals from one generation to the other guarantee that the achieved results from the previous run do not get lost.

The backcalculation of the moduli of a 23-layer pavement system showed promising future of the new PDGA algorithm in saving time and in achieving accurate results with less effort.

CHAPTER IX

SUMMARIES AND CONCLUSIONS

9.1 Summaries

The backcalculation of pavement moduli is very challenging since the multimodal nature of the search space reduces the ability of the classical backcalculation techniques in finding the global optima. In this book, the backcalculation of the pavement moduli using genetic algorithms was investigated thoroughly. The effect of the genetic operators and parameters on the performance of the backcalculation process was studied as well. In addition, the complexity of the search space was further studied using different backcalculation cases.

A new backcalculation program (BackGenetic3D) was developed based on the genetic algorithms theory that is capable of backcalculating the moduli of any pavement system with any number of pavement layers, loading configuration, and loading shape. The BackGenetic3D program employed new genetic operators and parameters such as the jump mutation, creep mutation, crossover, niching, tournament selection, and elitism. In addition, a superior random number generator (Mersenne Twister random Number Generator) was employed to ensure randomness in the generated population and numbers. The program was verified by analyzing a 3-layer pavement system using the BackGenetic3D program and other programs that use classical backcalculation techniques. The BackGenetic3D program was used to further backcalculate the pavement moduli of different cases and to study the many factors that affect the backcalculation process.

A new method was proposed to reduce the effort needed to optimize genetic parameters. In addition, the effects of the chromosome length, the fitness function, the number of layers, the number of sensors, and the range of the search space on the performance of the backcalculation process were investigated. The results of 3-, 4-, and 23-layer pavement systems were used to draw the conclusions and findings.

The automation of the genetic algorithms was investigated by reviewing the available adaptive methods in literature. A new population-sizing algorithm (PDGA) was developed that

196

eliminates the need for both the mutation probability and the crossover probability. The size of the population is determined by doubling the population size in steps while keeping the results of the previous population. The new algorithm was tested using a 23-layer pavement system.

9.2 Conclusions

Based on the research work performed in this book, the following conclusions can be drawn:

1) The backcalculation of the pavement moduli using classical backcalculation techniques can lead to misleading results due to the premature convergence, the local optima, and the need for seed moduli.

2) The use of the deflection RMSE error with classical backcalculation techniques can be misleading since different moduli sets can be obtained for the same RMSE error.

3) The least square error should not be used as the objective function when backcalculating the pavement moduli and the AREA method should be used instead.

4) Increasing the number of sensors can increase the accuracy of the backcalculated moduli.

5) Each unknown moduli should be presented by at least 15 bit strings.

6) The complexity of the search space depends on the number of sensors, the number of layers, ranges of moduli, fitness function, and the population size.

7) Small populations should not be used to backcalculate the pavement moduli and large populations should be used instead.

8) The sensitivity analysis can be used to reduce the number of trials to optimize the genetic parameters.

9) Niching and elitism should be used to enhance the performance of the genetic algorithms in multimodal functions.

10) The automation of the genetic algorithms can be done using the newly developed Dynamic Parameterless Genetic Algorithm (DPGA).

11) The BackGenetic3D program is an accurate and robust backcalculation program. The new program is capable of backcalculating the moduli of any pavement system with any arbitrary number of layers, loading configuration, and load distribution assuming elastic, linear, and homogeneous layers.

9.3 Future Studies

Based on the current work, some attractive problems could be further studied in the future. The backcalculation of the pavement moduli using field measurements from instruments installed inside the pavement should be considered. Many of the pavement sections in Ohio and in the US are instrumented with strain gages and hence the variation of the deflection within the pavement section and spatially can be obtained.

On the other hand, the backcalculation of the pavement using the actual load distribution from tires on the pavement sections should be considered. The BackGenetic3D program can be used to backcalculate the pavement moduli using the actual load configuration and distribution with data collected either along the surface of the pavement or within the pavement section.

In addition, the current program assumes elastic, linear, and homogeneous pavement layers. The nonlinearity of the stresses in the pavement layers should be considered in the backcalculation procedure especially under high surface loads. A comparison between backcalculated moduli assuming linear and nonlinear stress distributions should be carried out.

The current practice uses the surface deflections as the main inputs in the backcalculation process. This approach showed limitations during the backcalculation process where unique solutions are hard to find especially when the root mean square error is used as the objective function. Future research should focus more on developing better objective functions or on identifying other field measurements that can enhance the backcalculation process and the accuracy of the backcalculated moduli.

The location and number of sensors used in the backcalculation process should be further investigated. The current practice adopt the use of 6 to 9 sensors to measure the surface deflections placed at distances that are multipliers of loading plate radius. The effects of the distances between the sensors and the number of sensors should be further studied taking into account the load value and the pavement thickness.

The developed genetic algorithms in this book can be used with other advanced numerical simulation tools such as the finite element to backcalculate the pavement moduli.

REFERENCES

AASHTO Guide for Design of Pavement Structures (1993). American Association of State Highway and Transportation Officials.

Acum, W.A, Fox, L. (1951). "Computation of Load Stresses in a Three-Layer Elastic System", *Geotechnique*, Vol.2, pp. 293-300.

Akram, T., Scullion, T., and Smith, R.E. (1994). Comparing Laboratory and Backcalculated Layer Moduli on Instrumented Pavement Sections, Special Technical Publication (STP) 1198, ASTM, Philadelphia, PA.

Ancey, C. (2005). Notebook Introduction to Fluid Rheology, École Polytechnique Fédérale de Lausanne.

Anderson, M. (1988). Backcalculation of Composite Pavement Payer Moduli, PhD dissertation, The University of Kentucky, Kentucky.

Anderson, M. (1989). A Database Method for Back Calculation of Composite Pavement Layer Moduli, Nondestructive Testing of Pavements and Back Calculation of Moduli, ASTM STP 1026, A.J. Busch III and G.Y Baladi, eds., American Society of Testing and Materials, Philadelphia.

Arabas, J., Michalewicz, Z., and Mulawka, J. (1994). GAVaPS- A Genetic Algorithm with Varying Population Size, Proceeding of the 1st IEEE Conference on Evolutionary Computation, IEEE Press, Vol. 1, pp. 73-78.

Asphalt Institute (1991). Thickness Design: Asphalt Pavements for Highways and Streets, Manual Series No.1 (MS-1).

Bäck, T. (1992). The Interaction of Mutation Rate, Selection, and Self-adaptation within a Genetic Algorithm. In Manner, R. and Manderick, B., eds., Parallel Problem Solving from Nature, Amsterdam, North-Holland, pp. 85-94.

Bäck, T. (1993). Optimal Mutation Rates in Genetic Search, in Proceedings of the Fifth International Conference on Genetic Algorithms, Stephanie Forrest, ed., Morgan Kaufmann, pp. 2-8.

Bäck, T., and Schutz M. (1996). Intelligent Mutation Rate Control in Canonical Genetic Algorithm, Proceeding of the International Symposium on Methodologies for Intelligent Systems, pp. 158-167.

Bäck, T., Eiben, A.E., and Van der Vaart, N.A. (2000). An Empirical Study on GAs "without Parameters", In Schenauer, M., Deb, K., Rudolph, G., Yao, X., Lutton, E.,Merelo, J. J., and Schwefel, H-P. eds., Parallel Problem Solving from Nature PPSN V, Lecture Notes in Computer Science, Vol. 1917, pp. 315-324.

Bendana, L.J., Yang, W.S., and Lu, J. (1994). Interpreting Data from the Falling Weight Deflectometer, Engineering Research and Development Bureau, New York State Department of Transportation, Research Report 160.

Bentsen, R.A., Bush, A.J., and Harrison, J.A. (1989). Evaluation of Nondestructive Test Equipment for Airfield Pavements. Phase 1. Calibration Test Results and Field Data Collection, US Army Engineer Waterways Experiment Station, Washington, DC.

Bodare, A. and Orrje, O. (1985). Impulse Load on a Circular Surface in an Infinite Elastic Body-Closed Solution According to the Theory of Elasticity, Report No. 19, Department of Soil and Rock Mechanics, Royal Institute of Technology, Stockholm, Sweden.

Bonnaure, F., Gest, G., Gravois, A. and Uge, P. (1977). A New Method of Predicting The Stiffness of Asphalt Paving Mixtures, Proceedings of the Association of Asphalt Paving Technologists, No. 46, pp. 64–100.

Bonnaure, F., Gravois, A., and Udron, J. (1980). "A New Method of Pedicting Te Ftigue Lfe of Btuminous Mxes." *Journal of the Association of Asphalt Paving Technologist*, 49, pp. 499-529.

Boussinesq, J. (1876). "Essai theorique sur l'equilibre d'elasticite des massifs pulverulents compare a celui de massifs solides et sur la poussee des terres sans cohesion". *Memoires des Savants Etrangers*, Academie de Belgique, 40, Brussels.

Boussinesq, J. (1885). Application des potentials a' l'etude de l'equilibre et du movement des solides elastique, Gauthier-Villard, Paris.

Burmister, D.M. (1943). The Theory of Stresses and Displacements in Layered Systems and Applications to the Design of Airport Runways, Proceedings, Highway Research Board, Vol.23, pp. 126-144.

Burmister, D.M. (1945)."The General Theory of Stresses and Displacements in Layered Soil Systems", *Journal of Applied Physics*, Vol. 16, pps.89-94, 126-127, 296-302.

Bush, A.J. (1980). Nondestructive Testing for Light Aircraft Pavements, Phase II, Development of the Nondestructive Evaluation Methodology, Report No. FAA-RD-80-9-II, US Army Engineer Waterways Experiment Station, Washington, DC.

Bush, A. J. (1985). Computer Program BISDEF, United States Army, Corps of Engineers, Waterways Experiment Station.

Bush, A.J., and Alexander, D.R. (1985). Pavement Evaluation Using Deflection Basin Measurements and Layered Theory, Transportation Research Record 1022, Transportation Research Board, Washington, DC, pp. 16-29.

Chapra, S, and Canale, R. (1988). Numerical Methods for Engineers. McGraw Hill Book Co.

Cho, Y.S., and Lin, F.B. (2001). "Spectral Analysis of Surface Wave Response of Multilayer Thin Cement Mortar Slab Structures with Finite Thickness." *NDT and E International*, 34, pp. 115–122.

Chou, Y. J., and Lytton, R.L. (1991). Accuracy and Consistency of Backcalculated Pavement Layer Moduli, Transportation Research Record 1022, TRB, National Research Council, Washington, D.C., pp. 1-7.

COST 333 (1999). Development of New Bituminous Pavement Design Method, Final Report of the Action, European Commission, Brussels, Luxembourg.

Cramer, N.L. (1985). A Representation for The Adaptive Generation of Simple Sequential Programs, Proceedings of an International Conference on Genetic Algorithms and the Applications, Grefenstette, John J., ed., Carnegie Mellon University.

Daleiden, J.F., Killingsworth, B.M., Simpson, A.L., and Zamora, R.A. (1994). Analysis of procedures for establishing in situ subgrade moduli, Transportation Research Record 1462, TRB, National Research Council, Washington, D.C., pp. 102-107.

Danish Road Institute (2001). Verification of Flexible Pavement Response From a Field Test Report, Report 121.

Daridi, F., Kharma, N., and Salik, J. (2004). "Parameterless Genetic Algorithms: Review and Innovation", *IEEE Canadian Review*, No. 47, pp. 19-23.

Darwin, C. R. (1859). On The Origin of Species by Means of Natural Selection, or The Preservation of Favored Races in The Struggle for Life. First edition. London: John Murray.

Deb, K. (1995). Optimization for Engineering Design, Prentice-Hall, New Delhi.

Deb, K. (2005). Practical Optimization Using Evolutionary Methods. Kanpur Genetic Algorithm Lab (KanGAL), GAL Report No. 2005008, Kanpur, India.

De Jong, K. A. (1975). An Analysis of The Behavior of a Class of Genetic Adaptive Systems. PhDl dissertation, University of Michigan, Ann Arbor. (University Microfilms No. 76-9381).

De Jong, K. A. (1993). Genetic Algorithms are Not Function Optimizers. In L.D. Whitley, ed., Foundations of Genetic Algorithms 2. Morgan Kaufmann, pp. 5-17.

Eiben, A., Hinterding, R. and Michalewicz, Z. (1999). "Parameter Control in Evolutionary Algorithms". *IEEE Transactions on Evolutionary Computation*, Vol. 3 No. 2, 124-141.

Falkenauer, E. (1998). Genetic Algorithms and Grouping Problems, John Wiley and Sons.

203

Federal Highway Administration (FHWA) (1998). Techniques for Pavement Rehabilitation, Course No. FHWA-NHI-131008.

Federal Highway Administration (FHWA) (1990). Spring Load Restrictions, Pavement Newsletter No. 17, US Department of Transportation.

Fogarty, T., and Terence, C. (1989). Varying the Probability of Mutation in The Genetic Algorithm, Proceeding of the Third International Conference on Genetic algorithms, pp. 104-109.

Fogel, L.J., Owens, A.J., and Walsh, M.J. (1966). Artificial Intelligence through Simulated Evolution. Wiley, New York.

Foster, C.R., and Ahlvin, R.G. (1954). Stresses and Deflections Induced by a Uniform Circular Load, Proceedings, Highway Research Board, Vol.33, pp. 467-470.

Futuyma, Douglas J. (2005). Evolution. Sunderland, Sinauer Associates, Massachusetts.

Fwa, T.F., Tan, C.Y. and Chan, W.T. (1997). Backcalculation Analysis of Pavement Layer Moduli using Genetic Algorithms, Transportation Research Record 1905, TRB, National Research Council, Washington, D.C., pp 134–142.

Fwa, T.F., and Rani, T.S. (2005). Seed Modulus Generation Algorithm for Backcalculation of Flexible Pavement Moduli, Transportation Research Record 1905, TRB, National Research Council, Washington, D.C., 117-127.

Goldberg, D. E. (1987). Simple Genetic Algorithms and The Minimal, Deceptive Problem. In Davis, L., ed., Genetic Algorithms and Simulated Annealing, Los Altos, CA, Morgan Kaufmann, pp. 74–88.

Goldberg, D. E. (1989). Genetic Algorithms in Search, Optimization, and Machine Learning. Reading, M.A, Addison-Wesley.

Goldberg, D.E. (1993). "Making Genetic Algorithms Fly. A lesson from The Wright Brothers". *Advanced Technology for Developers*, No. 2, pp. 1-8.

Goldberg, D.E. (2002). The Design of Innovation, Lessons from and for Competent Genetic Algorithms. Kluwer Academic Publishers, Boston, MA.

Goldberg, D. E. and Deb, K. (1991). A Comparative Analysis of Selection Schemes used in Genetic Algorithms, Foundations of Genetic Algorithms, G. J. E. Rawlins, ed., pp. 69-93.

Goldberg, D. E., Deb, K., and Clark, J. H. (1992). "Genetic Algorithms, Noise, and the Sizing of Populations". *Complex Systems*, Vol. 6, pp. 333-362.

Goldberg, D.E., and Richardson, J. (1987). Genetic Algorithms with Sharing for Multimodal Function Optimization. Genetic algorithms and Their Applications: Proceedings of the Second International Conference on Genetic Algorithms, p. 41-49.

George, K.P, (2003). Falling Weight Deflectometer for Estimating Subgrade Moduli, Mississippi Department of Transportation, Report No. MS-DOT-RD-03-153.

Grefenstette, J.J. (1986). "Optimization of Control Parameters for Genetic Algorithms". *IEEE Transactions on Systems, Man, and Cybernetics*, Vol. 16, No. 1, p. 122-128.

Gustafson, M. (2004). An Analysis of Diversity in Genetic Programming. PhD dissertation, The University of Nottingham, United Kingdom,.

Haldane, J.B. (1953). The Measurement of Natural Selection. Proceedings of the 9th International Congress of Genetics, Vol. 1, pp. 480–487.

Harichandran, R., Mahmood, T., Raab, A., and Baladi, G. (1993). Modified Newton Algorithm for Backcalculation of Pavement Layer Properties, Transportation Research Record 1384, TRB, National Research Council, Washington, D.C., pp. 15-22.

Harichandran, R. S., Ramon, C. M., and Baladi, G. Y. (1994). MICHBACK User's Manual, Department of Civil and Environmental Engineering, Michigan State University, East Lansing, Michigan.

Harik, G. R., Cantú-Paz, E., Goldberg, D. E., and Miller, B. (1997). The Gambler's Ruin Problem, Genetic Algorithms, and The Sizing of Populations, Bäck, T. ed., Proceedings of 1997 IEEE International Conference on Evolutionary Computation, New York, IEEE Press, pp. 7–12.

Harik, G., and Lobo, F. (1999). A Parameter-less Genetic Algorithm, Banzhaf, W., Daida, J., Eiben, A. E., Garzon, M. H., Honavar, V., Jakiela, M., & Smith, R. E., eds., GECCO-99: Proceedings of the Genetic and Evolutionary Computation Conference, San Francisco, CA, Morgan Kaufmann, pp. 258–267.

Hesser, J. and Manner, R. (1991). Towards an Optimal Mutation Probability in Genetic Algorithms, Proceeding of the 1st Parallel Problem Solving from Nature, Springer, pp. 23-32.

Heukelom, W., and Foster, C. (1960). "Dynamic Testing of Pavements". *ASCE SM Journal*, Vol. 86, No. 1,pp. 1-28.

Hinterding, R., Michalewicz, Z., and Peachy, T. (1996). Self-Adaptive Genetic Algorithm for Numeric Functions, Proceeding of the Fourth International Conference on Parallel Problem Solving from Nature, in Lecture Notes from Computer Science, Springer Verlag, pp. 420-429.

Hoffman, M.S., and Thompson, M.R. (1981). Mechanistic Interpretation of Nondestructive Testing Deflections, Civil Engineering Studies, Transportation Engineering Series No. 32, Illinois Cooperative Highway and Transportation Research Program Series No. 190, University of Illinois, Urbana, IL.

Hoffman, M.S. (1983). "Loading Mode Effects on Pavement Deflections". *Journal of Transportation Engineering*, Vol. 109, No. 5, pp. 61-668.

Holland, J. H. (1975). Adaptation in Natural and Artificial Systems. Ann Arbor, University of Michigan Press.

Holland, J.H. (1992). Adaptation in Natural and Artificial Systems, An Introductory Analysis with Applications to Biology, Control and Artificial Intelligence, MIT Press.

Holtz, R., and Kovacs, W. (1981). An Introduction to Geotechnical Engineering, Prentice Hall, Inc., New Jersy.

Huang, Y. H. (1993). Pavement Analysis and Design, Prentice Hall, New Jersey.

Ioannides, A.M., Barenberg, E.J., and Lay, J.A. (1989). Interpretation of Falling Weight Deflectometer Results Using Principles of Dimensional Analysis, Proceedings of the Fourth International Conference on Concrete Pavement Design and Rehabilitation, Purdue University.

Irwin, L.H. (1983). User's Guide to MODCOMP2, Report No. 83-8, Cornell University Local Roads Program, Cornell University.

Irwin, L.H. (1992). Instructional Guide for Back-Calculation and the Use of MODCOMP 3, Version 3.6, Report 92-5, Cornell Local Roads Program, Cornell University.

Jones, R. (1963). "Following Changes in the Properties of Road Bases and Subbases by the Surface Wave Propagation Method". *Civil Engineering and Public Works Review*, Vol. 58, pp.613, 615,617.

Jones, R. (1965). "Thickness and Quality of Cemented Surfacing and Bases Measuring by a Non-Destructive Surface Wave Method". *Civil Engineering and Public Works Review*, Vol. 60, pp.523, 525,527,529.

Jones, R., and Thrower, E. (1965). "Effect of Interfacial Contact on the Propagation of Flexural Waves along a Composite Plate", *J. Sound Vib.*, Vol. 2, No. 2, pp. 167-174.

Julstrom, B.A. (1995). What Have You Done for Me Lately? Adapting Operator Probabilities in a Steady-state Genetic Algorithm, Proceeding of the Sixth International Conference on Genetic Algorithms, pp. 81-87, Morgan Kufmann.

Kameyama, S., Himeno, K., Kasahara, A., and Maruyama, T. (1998). Backcalculation of Pavement Layer Moduli using Genetic Algorithms, 8th International Conference on Asphalt Pavements, University of Washington, Seattle, Washington, pp 1375–1385.

Kauffman, S. (1993). The Origins of Order: Self-origination and Selection in Evolution, Oxford University Press, New York, NY.

Koza, J.R. (1992). Genetic Programming: On the Programming of Computers by Means of Natural Selection, MIT Press.

Lande, R., Arnold, S. (1983). "The Measurement of Selection on Correlated Characters". *Evolution*, Vol. 37, pp. 1210–1226.

Lobo, F. G. (2000). The Parameter-less Genetic Algorithm: Rational and Automated Parameter Selection for Simplified Genetic Algorithm Operation. Doctoral dissertation, Universidade Nova de Lisboa, Lisboa.

Lytton, R. L., Germann, F.P., Chou, Y.J., and Stoffels, S.M. (1990). Determining Asphaltic Concrete Pavement Structural Properties by Nondestructive Testing, NCHRP Report 327, Transportation Research Board.

Lytton, R.L., Roberts, F.L., and Stoffels, S. (1986). Determination of Asphaltic Concrete Pavement Structural Properties by Nondestructive Testing, Report NCHRP 10-27, Phase I, Texas Transportation Institute, Transportation Research Board, Washington, DC.

Mamlouk, M.S. (1985). Use of Dynamic Analysis in Predicting Field Multilayer Pavement Moduli, Transportation Research Record 1043, Transportation Research Board, Washington, DC, pp. 113-120.

Martincek, G. (1994). Dynamics of Pavement Structures, E & FN SPON, London.

Matsumoto, M. and Nishimura, T. (1998). "Mersenne Twister: a 623-dimensionally Equidistributed Uniform Pseudorandom Number Generator". *ACM Transactions on Modeling and Computer Simulation*, Vol. 8, No. 1, pp. 3-30.

McKay, R. (2000). Fitness Sharing in Genetic Programming. Proceedings of the Genetic and Evolutionary Computation Conference, Whitley, D. et al., eds., Las Vegas, NV, ,Morgan Kaufmann, pp. 435-442.

Michelow, J. (1963). Analysis of Stresses and Displacements in N-Layered Elastic System Under a Load Uniformly Distributed on a Circular Area, Chevron Oil Research, Richmond, California.

Minqiang, L., and Jisong, K. (2001). "A New Non-monotone Fitness Scaling for Genetic Algorithm"., Progress in Natural Science, Vol. 11, No. 8, pp. 622-630.

Mitchell, M. (1999). An Introduction to Genetic Algorithms, MIT Press.

Misra, D. and Sen, B. (1975). "Stresses and Displacements in Granular Materials due to Surface Load". International Journal of Engineering Science, Vol. 13, pp. 743-761.

Mühlenbein, H. (1991). Evolution in Time and Space-the Parallel Genetic Algorithm, Foundations of Genetic Algorithms, G. Rawlins, ed, Sand Mateo, Morgan Kaufman, pp. 316-338.

Mühlenbein, H. (1992). How Genetic Algorithms Really Work: Mutation and Hillclimbing, Parallel Problem Solving from Nature, Manner, R. and Manderick, B. eds., Amsterdam, North-Holland, pp. 15-26.

National Research Council (1993). SHRP's Layer Moduli Back Calculation Procedure, Strategic Highway Research Program, Report SHRP-P-655.

Nazarian, S., Yuan, D., and Baker, M.R. (1995). Rapid Determination of Pavement Moduli with Spectral-Analysis-of-Surface-Waves Method, Research Record 1243-1F, Center for Geotechnical and Highway Materials Research, University of Texas at El Paso.

Newmark, N.M. (1935). Simplified Computation of Vertical Pressures in Elastic Foundations, University of Illinois Engineering Experiment Station Circular 24, Urbana, Illinois.

Ochoa, G., Harvey, I. and Buxton, H. (1999). On Recombination and Optimal Mutation Rates, Proceedings of Genetic and Evolutionary Computation Conference (GECCO-99).

Pan, E. (1989a). "Static Response of a Transversely Isotropic and Layered Half-space to General Dislocation Sources". *Phys. Earth Planet. Inter.,* Vol. 58, pp. 103-117.

Pan, E. (1989b). "Static Response of a Transversely Isotropic and Layered Half-space to General Surface Loads". *Phys. Earth Planet. Inter.,* Vol. 54, pp. 353-363.

Pan, E. (1990). "Thermoelastic Deformation of a Transversely Isotropic and Layered Half-space by Surface Loads and Internal Sources". *Phys. Earth Planet. Inter.,* Vol. 60, pp. 254-264.

Pan, E. (1997). "Static Green's Functions in Multilayered Half-spaces". *Applied Mathematical Modelling,* Vol. 21, pp. 509-521.

Pan, E. and Alkasawneh, W. (2006). An Exploratory Study on Functionally Graded Materials with Application to Multilayered Pavement Design, Report No. FHWA/OH- 6936-01, Ohio Department of Transportation, Columbus, Ohio.

211

Potter, D.W, and Donald, G.S. (1985). Revision of The NAASRA Interim Guide to Pavement Thickness Design, Australian Road Research, Technical Note 2, Vol. 15, No. 2.

Ray, T.S. (1991). Is It Alive or Is It GA, Proceedings of the Fourth International Conference on Genetic Algorithms, Morgan Kaufmann, pp. 527-534.

Rechenberg, I. (1965). Cybernetic Solution Path of an Experimental Problem, Ministry of Aviation, Royal Aircraft Establishment, United Kingdom.

Rechenberg, I. (1973). "Evolutions Strategie: Optimierung Technischer Systeme Nach Prinzipien der Biologischen Evolution". Frommann-Holzboog, Stuttgart.

Reddy, M.A., Reddy, K.S., and Pandey, B.B. (2004). "Selection of Genetic Algorithm Parameters for Backcalculation of Pavement Moduli". *The International Journal of Pavement Engineering*, Vol. 5, No. 2, pp. 81-90.

Roesset, J.M., and Shao, K.Y. (1985). Dynamic Interpretation of Dynaflect and Falling Weight Deflectometer Tests, Transportation Research Record 1022, Transportation research Board, National Research Council, Washington, DC, pp. 7-16.

Rosca, J. and Ballard, D. (1995). Causality in Genetic Programming, Proceedings of the Sixth International Conference on Genetic Algorithms, In Eshelman, L., ed., Pittsburgh, PA, Morgan Kaufmann, pp. 256-263.

Ryan, C. (1994). Pygmies and Civil Servants, Advances in Genetic Programming, Kinnear, Jr., K., ed., MIT Press, Cambridge, MA, pp. 243-263..

Schaffer, J. D., Caruana, R. A., Eshelman, L. J., and Das, R. (1989). A Study of Control Parameters Affecting Online Performance of Genetic Algorithms for Function Optimization, Proceedings of the Third International Conference on Genetic Algorithms Schaffer, J. D. ed., San Mateo, CA, Morgan Kaufmann, pp. 51–60.

Schiffman, R. L. (1962). General Solution of Stresses and Displacements in Layered Elastic Systems, Proceedings of the International Conference on the Structural Design of Asphalt Pavement, University of Michigan, Ann Arbor, USA.

Schlierkamp-Voosen, D., and Muhlenbein, H. (1996). Adaptation of Population Sizes by Competing Subpopulations, Proceeding of International Conference on Evolutionary Computation (ICEC'96), Negoya, Japan, pp. 330-335.

Schwefel, H-P. (1977). "Numerische Optimierung von Computer-modellen Mittels der Evolutionsstrategie". Basel, Birkaeuser.

Seeds, S.B., Ott, W.C., Milchail, M., and Mactutis, J.A. (2000). Evaluation of Laboratory Determined and Nondestructive Test Based Resilient Modulus Values From WesTrack Experiment, Nondestructive Testing of Pavements and Backcalculation of Moduli, ASTM STP 1375, S. Tayabji and O. Lukanen, eds., ASTM.

Shell International Petroleum Company Limited (1978). Shell Pavement Design Manual.

Shook, J. F., Finn, F. N., Witczak, M. W., and Monismith, C. L. (1982). Thickness Design of Asphalt Pavements – The Asphalt Institute Method. Fifth International Conference on the Structural Design of Asphalt Pavements, Vol. 1, pp. 17-44.

Sivaneswaran, N., Kramer, S.L., and Mahoney, J.P. (1991). Advanced Backcalculation using a Nonlinear Least Squares Optimization Technique, Presented at the 70th Annual Meeting of The Transportation Research Board, Washington, D.C.

Skok, E.L., Clyne, T.R., Johnson, E., Timm, D.H., and Brown, M.L. (2003). Best Practices for The Design and Construction of Low Volume Roads Revised, Report No. MN/RC-2002-17REV, Minnesota Department of Transportation Research Services Section, Minnesota.

Smith, S.F. (1980). A Learning System Based on Genetic Adaptive Algorithms, PhD dissertation, University of Pittsburgh.

Smith, R.E. (1993). Adaptively Resizing Populations: An Algorithm and Analysis, Technical Report No. 93001, The University of Alabama, Tuscaloosa, AL.

Smith, R. E. and Smuda, E. (1995). "Adaptively Resizing Populations: Algorithm, Analysis, and First Results". *Complex Systems*, Vol. 9, pp. 47–72.

Spears, W.M. and De Jong, K.A. (1991). On the Virtues of Parameterized Uniform Crossover, Proceeding of the Fourth International Conference on Genetic Algorithms, R.K. Belew and L.B. Booker, eds., Morgan Kaufmann, pp. 230-236.

Srinivas, M., and Patniak, L. M. (1994). "Adaptive Probabilities of Crossover and Mutation in Genetic Algorithms". *IEEE Transactions on Systems, Man and Cybernetics,* Vol. 24, No. 4, pp. 17-26.

Stolle, D. (2002). "Pavement Displacement Sensitivity to Layer Moduli." *Canadian Geotechnical Journal*, Vol. 39, pp. 1395-1398.

Syswerda, G. (1989). Uniform Crossover in Genetic Algorithms, Proc. of the 3rd Int. Conf. on Genetic Algorithms, Schaffer, J. D., ed., Morgan Kaufmann, pp. 2-9.

Syswerda, G. (1991). Schedule Optimization using Genetic Algorithms, Handbook of Genetic Algorithms, Lawrence Davis, ed., Van Nostrand Reinhold, New York, pp. 332-349.

Szendrei, M., and Freeme, C. (1970). Dynamic Techniques for Testing Pavement Structures, National Institute of Road Research, Bulletin 9, South Africa.

Tsai, B., Kannekanti, V., and Harvey, J.T. (2004). Application of Genetic Algorithm in Asphalt Pavement Design, Transportation Research Board, No. 1891, National Research Council, Washington, D.C., pp. 112-120.

Ullidtz, P. (1973). The Use of Dynamic Plate Loading Tests in Design of Overlays, The Conference on Road Engineering in Asia and Australia, Kuala Lumpur.

Ullidtz, P. (1987). Pavement Analysis, Developments in Civil Engineering, Elsevier, Amestrdam.

Ullitdz, P. (1998). Stresses and Strains in a Two-Dimensional Particulate Material, 74th Annual Meeting of the Transportation Research Board, Washington D.C.

Uzan, J. (1986). Computer Program MODULUS, Appendix E, NCHRP Project 10-27, Texas Transportation Institute, Texas A&M University System, College Station, Texas, pp. E1-E22.

Uzan, J., Lytton, R.L., and Germann, F.P. (1989). General Procedure for Backcalculating Layer Moduli, Nondestructive Testing of Pavements and Backcalculation of Moduli, ASTM STP 1026, A.J. Buch III and G.Y. Baladi, eds., American Society for Testing and Materials, pp. 217-228.

Uzan, J., Michalek, C., Paredes, M., and Lytton, R. (1988). A Microcomputer Based Procedure for Backcalculating Layer Moduli from FWD Data, Research Report 113-1, Texas Transportation Institute.

Van der Loo, J.M. (1982). Simplified Method for Evaluation of Asphalt Pavements, Proceedings the Fifth International Conference on Structural Design of Asphalt Pavements, The Netherlands, Vol. 1, pp. 475-481.

Von Quintus, H.L. and Killingsworth, B. (1997). Design Pamphlet for the Backcalculation of Pavement Layer Moduli in Support of the 1993 AASHTO Guide for the Design of Pavement Structures, Publication No. FHWA-RD-97-076, Washington, DC, Federal Highway Administration.

Von Quintus, H.L., and Simpson, A. (2002). Back-Calculation of Layer Parameters for LTPP Test Sections Volume II: Layered Elastic Analysis for Flexible and Rigid Pavements, Report No. FHWA-RD-01-113, Federal Highway Administration.

Warren, T., and Diekmann, W.L. (1963). Numerical Computation of Stresses and Strains in Multilayer Asphalt Pavement System, Internal Report, Chevron Research Company.

Westergaard, H.M. (1938). A Problem of Elasticity Suggested by a Problem in Soil Mechanics: A Soft Material Reinforced by Numerous Strong Horizontal Sheets, Contributions to the Mechanics of Solids, Stephen Timoshenko 60th Anniversary Volume, Macmillian, New York, pp. 268-277.

White, A. (1993). Integrating Automata with Genetic Algorithms in Order to Provide Adaptive Operators, M.S. Thesis, Ottawa-Carleton Institute for Computer Science, Carleton University, Ottawa, Ontario.

Whitley, D. (1994). "A Genetic Algorithm Tutorial". *Statistics and Computing Journal*, Vol. 4, No. 2, pp. 65-85.

Winston, P. (1992). Artificial Intelligence, Third Edition. Addison-Wesley.

Yoder, E.J. (1959). Principles of Pavement Design, John Wiley & Sons, Inc., London.